Number Fluency

Peter Clarke

Developing mental fluency in numerical skills

Year 6

Acknowledgements

The author wishes to thank Brian Molyneaux and Mike Askew
for their valuable contribution to this publication.

Published by BEAM at Nelson Thornes
Delta Place
27 Bath Road
Cheltenham GL53 7TH

Telephone 01242 267 287
Fax 01242 253 695
Email cservices@nelsonthornes.com

ISBN 978 1 9062 2485 1

British Library Cataloguing-in-Publication Data
Data available

Cover photo: Dreamstime

Printed in Croatia by Zrinski

14 13 12 11 10 / 9 8 7 6 5 4 3 2 1

Contents

Introduction

The BEAM *Number Fluency* series is a set of six books, one for each year from Year 1 to Year 6, that aims to support the development of number fluency in basic numerical skills through individual and paired activities.

The books comprise teaching strategies plus photocopiable activity resources that, for ease of teaching and photocopy reproduction, are also available online at no extra cost to book purchasers (see page 9).

Each book covers key learning objectives based on the Primary Numeracy Strategy (PNS): *Primary Framework for Mathematics* (2006). These objectives are organised into six sections, each of which addresses a key aspect of becoming fluent in number.

The six sections are the same across all six books and are:

1. Comparing and ordering numbers, including decimals

2. Place value and partitioning

3. Understanding and using tens

4. Deriving and recalling addition and subtraction facts, and using that knowledge

5. Deriving and recalling multiplication and division facts, and using that knowledge

6. Mental calculation methods

The chart on page 10 links each of these six sections to the relevant Year 6 strand objective and planning blocks and units from the PNS *Primary Framework for Mathematics* (2006). Refer to this chart when choosing a section.

Children develop fluency in number through a combination of four key elements:

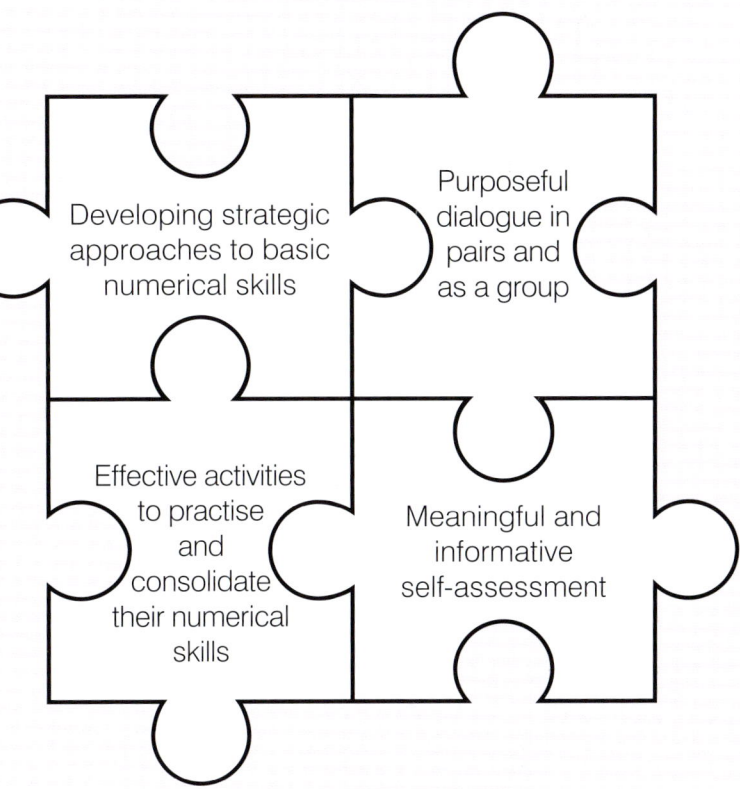

 **Content**

Each section begins with an introduction that includes:

- strategic approaches to develop fluency skills

- a brief description of the individual and paired activities

- teaching suggestions for ways to help children develop the fluency skills and provide further practice

- an individual child self-assessment record

For a more detailed explanation of each of these features, refer to pages 8 and 9.

Each of the six sections is divided into two levels: Level 1 (easier) and Level 2 (harder). There are two types of activities at each level: an individual activity and a paired activity. Both require children to work in pairs.

The individual activity contains two worksheets: A and B. Child A uses worksheet A; Child B uses worksheet B. The children work individually, but each worksheet contains the answers to their partner's questions. This enables each child to correct their partner's answers and to discuss the results.

For each of the paired activities, there are two worksheets, A and B. Child A uses worksheet A, and Child B uses worksheet B. Here, the children need to work together to complete their worksheets. These activities are either self-checking or pair-checking.

The diagram below aims to explain the structure of *Number Fluency*, using Year 6 Section 1 as an example.

Number Fluency Year 6

Section 1

Comparing and ordering numbers, including decimals

PNS *Framework for Mathematics* (2006) objective:

- Use decimal notation for tenths, hundredths and thousandths; partition, round and order decimals with up to three places and position them on a number line

---

Number Fluency Level 1 objective:

- Order decimals with up to three places and position them on a number line [some support]

Number Fluency Level 2 objective:

- Order decimals with up to three places and position them on a number line [less support]

Individual activity 1A	Paired activity 1A
Individual activity 1B	Paired activity 1B

Individual activity 2A	Paired activity 2A
Individual activity 2B	Paired activity 2B

The chart on page 11 shows Level 1 and Level 2 objective coverage for each of the six sections in *Number Fluency* Year 6. Refer to this chart to differentiate not only for particular individuals and pairs of children but also when choosing the level of work that is most suitable for a specific class, as well as the time of year in which you are teaching or consolidating the objective.

Suggestions for using *Number Fluency*

Number Fluency is a flexible resource that you can use in many different ways. One suggestion includes:

> Decide which objective you wish to develop fluency in (refer to the appropriate section).

↓

> Choose the appropriate level (Level 1 or 2).

↓

> Provide each pair with the individual activity to gain some self-assessed awareness of their level of fluency (Child A using worksheet A and Child B using worksheet B).

↓

> Children work together on the paired activity (A and B) with the explicit intention of supporting each other to improve their fluency.

↓

> Children repeat the individual activity to check their progress (this time, Child A using worksheet B and Child B using worksheet A).

As well as the individual activities and paired activities, *Number Fluency* offers:

- strategic approaches to develop fluency
- further activities to develop fluency.

Use these suggestions as you see fit:

- before children complete the first individual activity
- after children complete the first individual activity and before they do the paired activity
- after children complete the paired activity and before they do the second individual activity
- after children complete the second individual activity.

Number Fluency also provides an individual child self-assessment record for each section. Encourage the children to use these records to monitor their own progress and to see their improvement against their own baseline as opposed to being compared against other class members.

 ## The importance of promoting effective speaking and listening in developing children's mathematical understanding

Number Fluency recognises the importance of getting children to work collaboratively. By working as a pair, children learn from each other, confirming their mathematical knowledge and identifying for themselves, in a non-threatening environment, any misconceptions they may hold.

The recommendations in both the *Independent Review of Mathematics Teaching in Early Years Settings and Primary Schools* (2008) and the *Independent Review of the Primary Curriculum* (2008) reported on the importance of actively promoting speaking and listening in mathematics.

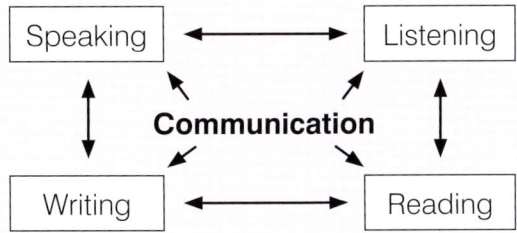

Each of the planning blocks and units in the PNS *Primary Framework for Mathematics* (2006) shows links with the relevant objectives from PNS *Speaking, Listening, Learning: Working with children in Key Stages 1 and 2* (2003). All of the activities in *Number Fluency* Year 6 aim to cover each of the following Speaking, Listening, Learning Year 6 objectives:

Speaking

- Use a range of oral techniques to present persuasive arguments and engaging narratives
- Participate in whole-class debate using the conventions and language of debate, including Standard English
- Use techniques of dialogic talk to explore ideas, topics or issues

Listening

- Analyse and evaluate how speakers present points effectively through use of language, gesture, models and images
- Make notes when listening for a sustained period and discuss how note taking varies depending on context and purpose

Group discussion and interaction

- Understand and use a variety of ways to criticise constructively and respond to criticism
- Identify the ways spoken language varies according to differences in context and purpose of use

How to use this book

Each of the six sections in **Number Fluency** Year 6 includes two pages offering advice on how you can teach the objective strategically and how the individual and paired activities support this.

Strategic approaches to developing fluency

Offers several approaches as to how to teach the objective strategically. Use these activities as part of the main teaching activity before children complete an individual or paired activity.

Individual and paired activities

Description of each of the individual and paired activities explaining the purpose of the activity, and one or two questions that you might use to discuss the activity with the children

Level 1 Individual activity

Level 1 Paired activity

Level 2 Individual activity

Level 2 Paired activity

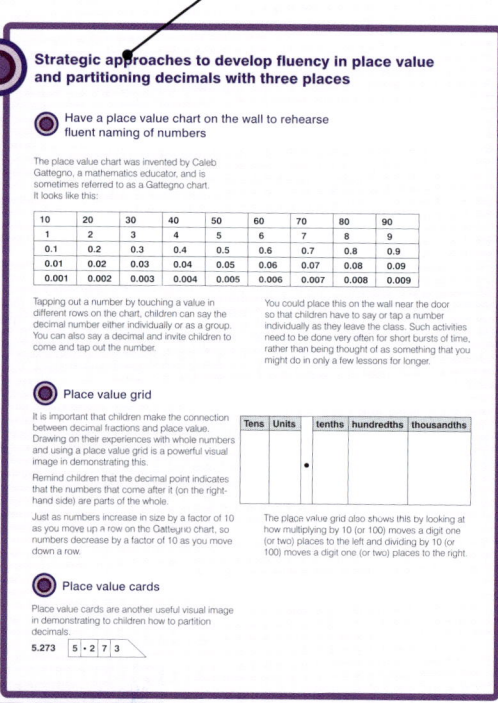

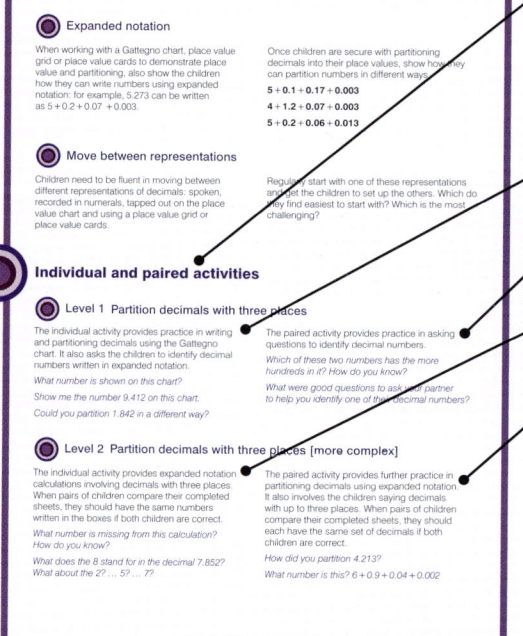

Each of the six sections in *Number Fluency* Year 6 also includes further activities to develop fluency and an individual pupil record sheet.

Further activities to develop fluency

Suggestion of further teaching activities to develop the fluency further. Use these activities either before or after the children complete the individual and paired activities.

My record sheet

Each section includes an individual child self-assessment sheet.

The top half of the sheet states the fluency objectives in child speak with 'smileys' (☺☺☹) to record self-assessment before and after the activities.

The bottom half of the sheet has three or four open-response questions to encourage the child to reflect on their learning.

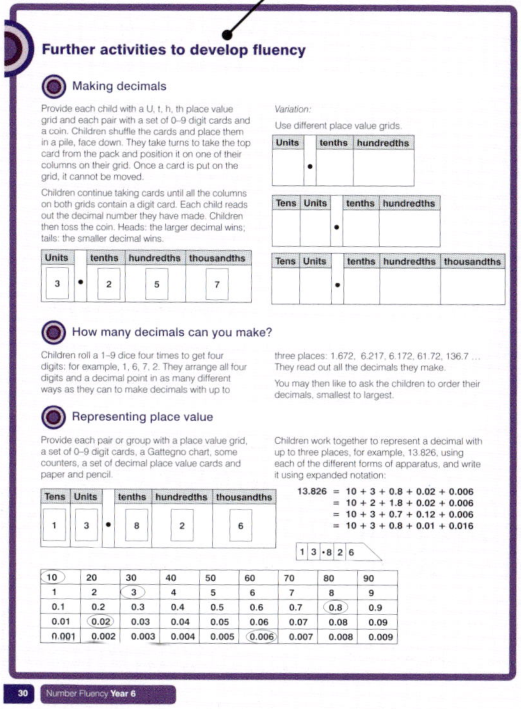

Further activities to develop fluency

Making decimals

Provide each child with a U, t, h, th place value grid and each pair with a set of 0–9 digit cards and a coin. Children shuffle the cards and place them in a pile, face down. They take turns to take the top card from the pack and position it on one of their columns on their grid. Once a card is put on the grid, it cannot be moved.

Children continue taking cards until all the columns on both grids contain a digit card. Each child reads out the decimal number they have made. Children then toss the coin. Heads: the larger decimal wins; tails: the smaller decimal wins.

Units		tenths	hundredths	thousandths
3	•	2	5	7

Variation:
Use different place value grids.

Units	tenths	hundredths
	•	

Tens	Units		tenths	hundredths
		•		

Tens	Units		tenths	hundredths	thousandths
		•			

How many decimals can you make?

Children roll a 1–9 dice four times to get four digits: for example, 1, 6, 7, 2. They arrange all four digits and a decimal point in as many different ways as they can to make decimals with up to three places: 1.672, 6.217, 6.172, 61.72, 136.7 … They read out all the decimals they make.

You may then like to ask the children to order their decimals, smallest to largest.

Representing place value

Provide each pair or group with a place value grid, a set of 0–9 digit cards, a Gattegno chart, some counters, a set of decimal place value cards and paper and pencil.

Tens	Units		tenths	hundredths	thousandths
1	3	•	8	2	6

Children work together to represent a decimal with up to three places, for example, 13.826, using each of the different forms of apparatus, and write it using expanded notation:

13.826 = 10 + 3 + 0.8 + 0.02 + 0.006
= 10 + 2 + 1.8 + 0.02 + 0.006
= 10 + 3 + 0.7 + 0.12 + 0.006
= 10 + 3 + 0.8 + 0.01 + 0.016

1	3	•	8	2	6

10	20	30	40	50	60	70	80	90
1	2	3	4	5	6	7	8	9
0.1	0.2	0.3	0.4	0.5	0.6	0.7	0.8	0.9
0.01	0.02	0.03	0.04	0.05	0.06	0.07	0.08	0.09
0.001	0.002	0.003	0.004	0.005	0.006	0.007	0.008	0.009

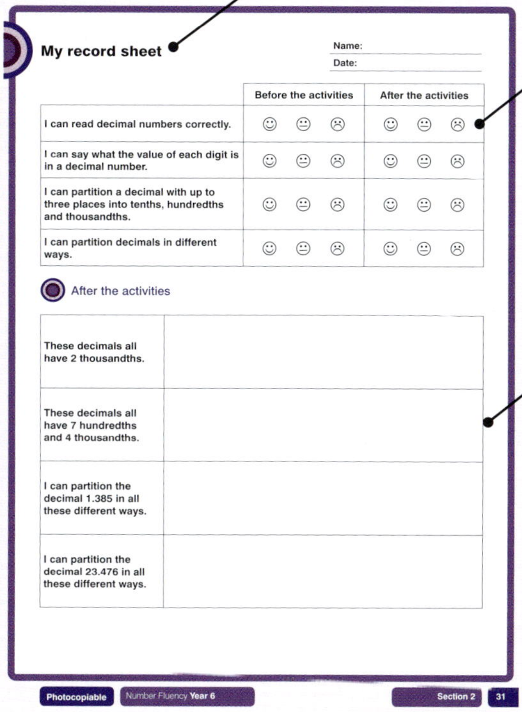

My record sheet

Name: _____
Date: _____

	Before the activities			After the activities		
I can read decimal numbers correctly.	☺	☺	☹	☺	☺	☹
I can say what the value of each digit is in a decimal number.	☺	☺	☹	☺	☺	☹
I can partition a decimal with up to three places into tenths, hundredths and thousandths.	☺	☺	☹	☺	☺	☹
I can partition decimals in different ways.	☺	☺	☹	☺	☺	☹

After the activities

These decimals all have 2 thousandths.	
These decimals all have 7 hundredths and 4 thousandths.	
I can partition the decimal 1.385 in all these different ways.	
I can partition the decimal 23.476 in all these different ways.	

Online access

Purchasers of the book can access the Record sheets for each section of each book, and all of the photocopiable Activity pages, online, for ease of photocopying or for interactive whiteboard display. To use these go to the following unique web address: www.beam.co.uk/numberfluency-FF9

Chart linking to the PNS *Primary Framework for Mathematics* (2006)

Number Fluency section	PNS *Primary Framework for Mathematics* (2006)		
	Strand	Objective	Planning block and unit
1. Comparing and ordering numbers, including decimals	2: Counting and understanding number	Use decimal notation for tenths, hundredths and thousandths; partition, round and order decimals with up to three places, and position them on the number line	A1, A2, A3
2. Place value and partitioning	2: Counting and understanding number	Use decimal notation for tenths, hundredths and thousandths; partition, round and order decimals with up to three places, and position them on the number line	A1, A2, A3
3. Understanding and using tens	3: Knowing and using number facts	**Use knowledge of place value and multiplication facts to 10 × 10 to derive related multiplication and division facts involving decimals: for example, 0.8 × 7, 4.8 ÷ 6**	A1, B1, E1, A2, B2, B3, E3
4. Deriving and recalling addition and subtraction facts, and using that knowledge	2: Counting and understanding number	Find the difference between a positive and a negative integer, or two negative integers, in context	A1
5. Deriving and recalling multiplication and division facts, and using that knowledge	3: Knowing and using number facts	Use knowledge of multiplication facts to derive quickly squares of numbers to 12 × 12 and the corresponding squares of multiples of 10	B1, B2, B3
6. Mental calculation methods	4: Calculating	Calculate mentally with integers and decimals: U.t ± U.t, TU × U, TU ÷ U, U.t × U, U.t ÷ U	A1, D1, A2, D2, A3, D3

Note: Key objectives are in **bold**.

Individual and paired activities

Number Fluency section	Level 1 Objective coverage	Pages		Level 2 Objective coverage	Pages	
		Individual activity 1A, 1B	Paired activity 1A, 1B		Individual activity 2A, 2B	Paired activity 2A, 2B
1. Comparing and ordering numbers, including decimals	Order decimals with up to three places, and position them on a number line [some support]	18, 19	20, 21	Order decimals with up to three places, and position them on a number line [less support]	22, 23	24, 25
2. Place value and partitioning	Partition decimals with three places	32, 33	34, 35	Partition decimals with three places [more complex]	36, 37	38, 39
3. Understanding and using tens	Use knowledge of place value and multiplication facts to 10×10 to derive multiplication and division facts involving decimals $U.t \times U$, $U.t \div U$ [restricted range of numbers]	46, 47	48, 49	Use knowledge of place value and multiplication facts to 10×10 to derive multiplication and division facts involving decimals $U.t \times U$, $U.t \div U$ [wider range of numbers]	50, 51	52, 53
4. Deriving and recalling addition and subtraction facts, and using that knowledge	Find the difference between a positive and a negative, or two negative integers, in context [restricted range of numbers]	60, 61	62, 63	Find the difference between a positive and a negative, or two negative integers, in context [wider range of numbers]	64, 65	66, 67
5. Deriving and recalling multiplication and division facts, and using that knowledge	Recall squares of numbers to 12×12	74, 75	76, 77	Derive squares of multiples of 10 to 120×120	78, 79	80, 81
6. Mental calculation methods	Calculate mentally with integers: specifically, $TU \pm TU$, $TU \times U$, $TU \div U$	88, 89	90, 91	Calculate mentally with decimals: specifically, $U.t \pm U.t$, $U.t \times U$, $U.t \div U$	92, 93	94, 95

Comparing and ordering numbers, including decimals

Level 1

- Order decimals with up to three places, and position them on a number line [some support]

Level 2

- Order decimals with up to three places, and position them on a number line [less support]

Strategic approaches to develop fluency in comparing and ordering decimals with up to three places

 ## Work on placing decimals on an empty number line

Children need to understand two main aspects of place value: the value of digits and knowing where a number lies in relation to other numbers.

There is a lot of research in psychology that shows that the brain stores numbers in a linear form. An empty number line helps children to comprehend the number system and think about the relative size of numbers. By using a number line with decimals, children's understanding of

the relationship between numbers will improve. It is important that they realise that markers on a number line can represent divisions other than one unit.

Present children with numbers that have different numbers of decimal places for ordering, to tackle the common misconception that the more digits there are after the decimal point, the bigger the number.

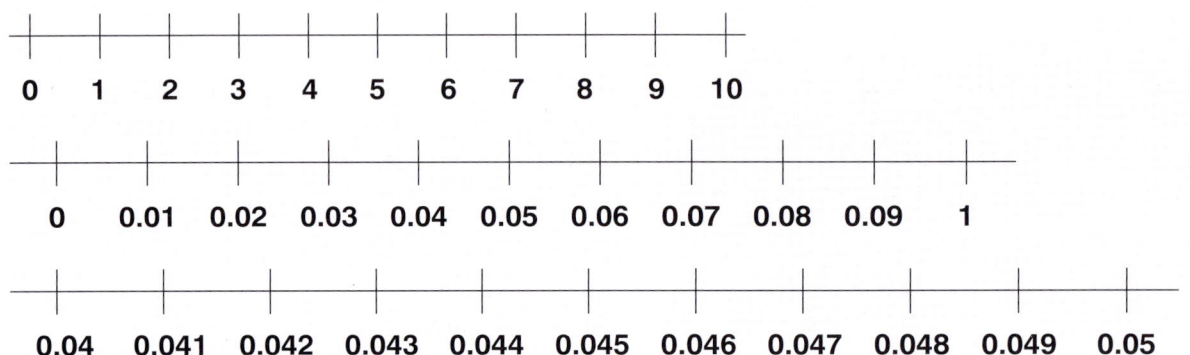

 ## Decimal number grid

For decimal numbers with up to two places, use a 0.01–1 grid so that each square represents 0.01 and each row represents 0.1. Discuss the similarities between this and a 1–100 grid and the effect of repeatedly adding the same decimal number: for example, 0.01 or 0.1.

0.01	0.02	0.03	0.04	0.05	0.06	0.07	0.08	0.09	0.1
0.11	0.12	0.13	0.14	0.15	0.16	0.17	0.18	0.19	0.2
0.21	0.22	0.23	0.24	0.25	0.26	0.27	0.28	0.29	0.3
0.31	0.32	0.33	0.34	0.35	0.36	0.37	0.38	0.39	0.4
0.41	0.42	0.43	0.44	0.45	0.46	0.47	0.48	0.49	0.5
0.51	0.52	0.53	0.54	0.55	0.56	0.57	0.58	0.59	0.6
0.61	0.62	0.63	0.64	0.65	0.66	0.67	0.68	0.69	0.7
0.71	0.72	0.73	0.74	0.75	0.76	0.77	0.78	0.79	0.8
0.81	0.82	0.83	0.84	0.85	0.86	0.87	0.88	0.89	0.9
0.91	0.92	0.93	0.94	0.95	0.96	0.97	0.98	0.99	1

Model decimals using a place value chart and place value cards

Write a decimal number such as 5.796 and use the Gattegno place value chart to discuss with the children the value of each of the digits (see Section 2).

10	20	30	40	50	60	70	80	90
1	2	3	4	5	6	7	8	9
0.1	0.2	0.3	0.4	0.5	0.6	0.7	0.8	0.9
0.01	0.02	0.03	0.04	0.05	0.06	0.07	0.08	0.09
0.001	0.002	0.003	0.004	0.005	0.006	0.007	0.008	0.009

Similarly, use place value cards.

5	•7	9	6

Focus on the vocabulary of decimals and encourage children to read decimal numbers, using the language of place value, including tenths, hundredths and thousandths, so that, for example, they know the number comprising five units, seven tenths, nine hundredths and sixth thousandths is written as 5.796.

Individual and paired activities

 ## Level 1 Order decimals with up to three places, and position them on a number line [some support]

The individual activity provides practice in comparing sets of decimals using the language of 'largest' and 'smallest', then ordering some of these decimals on partially numbered number lines.

Which of these decimals is the largest? How do you know that?

Where does this decimal belong on the number line?

The paired activity provides an opportunity for children to develop a mental image of where decimal numbers belong on a number line. When pairs have completed their sheets, they compare their number lines. If they are correct, they should have identical decimals on their number lines. If not, they need to discuss why.

Why did you write 1.339 here?

What decimal goes before/after this decimal?

 ## Level 2 Order decimals with up to three places, and position them on a number line [less support]

The individual activity provides practice in ordering sets of numbers with up to three decimal places. The children then write decimals and position them between some of these decimals to three places, so that the decimals are still in order.

Which is the largest/smallest number in this set of decimals? How do you know?

Tell me a decimal that lies between 1.415 and 1.54.

The paired activity provides practice in comparing two decimals using the vocabulary 'larger' and also an opportunity for children to write decimals that lie between two other decimals. The children then order a set of nine decimals. If they are correct, both children should have the same decimals written in the nine circles.

Which is larger 0.207 or 0.224? How do you know that?

Look at these two decimals. Tell me a decimal that is between these two numbers.

Which is the smallest of these decimals? Which is the largest?

Further activities to develop fluency

 ## Four in a row

Each pair has a 0.01–1 decimal number grid (100 squares), a 0 digit card, a set of 1–9 digit cards and a pile of counters in two different colours.

Children decide who has which colour counter. They shuffle the 1–9 digit cards and place them in a pile, face down. They take turns to choose two cards and put them to the right of the 0 digit card and the counter to make a decimal number such as 0.74.

They then put one of their coloured counters on that decimal on the number grid. When they have used all the cards, they reshuffle and continue. The winner is the first player to place four of their counters in a row, column or diagonal.

 ## Placing decimals on a number line

Children draw a large number line labelled 0.1 to 1 on a large sheet of paper.

They shuffle a set of 0–9 digit cards and place them in a pile, face down. They take turns to choose two cards and put them to the right of a counter to make a decimal number such as 0.43. They then write that decimal on the approximate position on the number line.

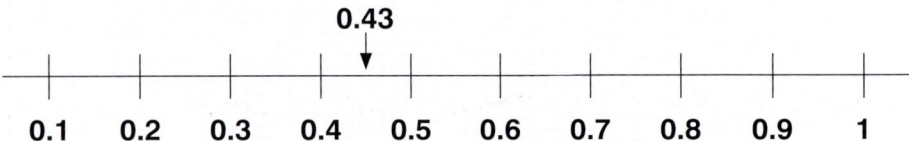

When they have used all the cards, they reshuffle and continue.

My record sheet

Name: _____

Date: _____

	Before the activities			After the activities		
I can compare two decimals with up to three places and say which is the larger.	☺	😐	☹	☺	😐	☹
I can order a set of decimals with up to three places.	☺	😐	☹	☺	😐	☹
I can place a set of decimals with up to three places on a number line.	☺	😐	☹	☺	😐	☹

◉ After the activities

Here are some decimals with three places in order, smallest to largest.

Here are some decimals with three places written on a number line.

Here are some decimals with two and three places written on a number line.

Individual activity 1A

Name:

Date:

Order decimals with up to three places, and position them on a number line [some support]

For each set of decimals:

• draw a circle around the largest decimal

• draw a box around the smallest decimal

1.117	1.109	1.119	1.127	1.11
0.039	0.04	0.42	1.113	1.111
0.045	0.11	1.11	0.043	1.115

0.04	1.122	1.125	0.05	0.044
1.129	0.045	0.054	1.13	0.113
1.3	1.112	1.113	1.131	1.12

Look at the decimals you circled or drew a box around.
Write them on the number lines.

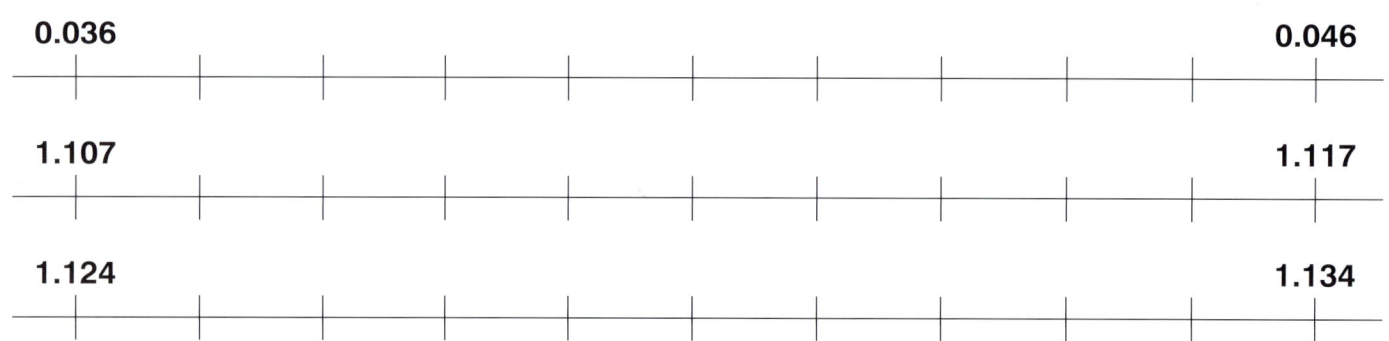

0.036   0.046

1.107   1.117

1.124   1.134

Ask your partner to check your number lines.

Talk with your partner about how you knew
where to write the decimals on the number lines.

_____

Answers to 1B

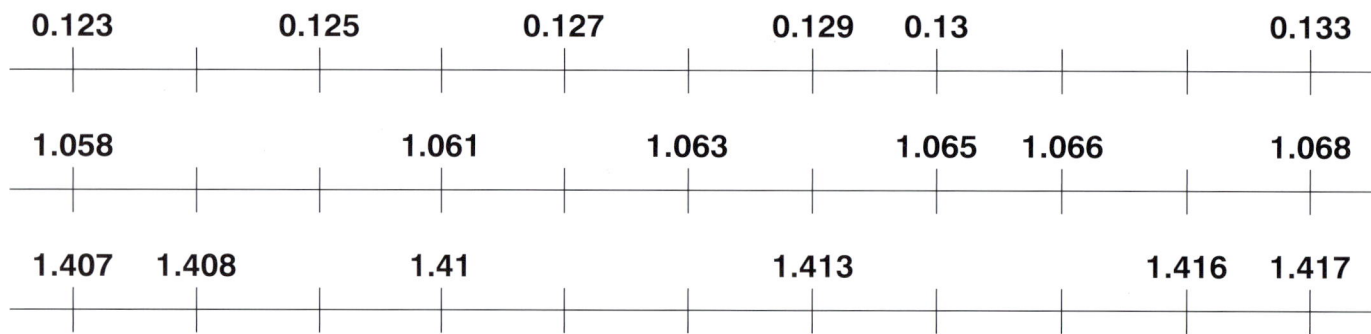

0.123   0.125   0.127   0.129   0.13   0.133

1.058   1.061   1.063   1.065   1.066   1.068

1.407   1.408   1.41   1.413   1.416   1.417

Individual activity 1B

Order decimals with up to three places, and position them on a number line [some support]

For each set of decimals:

• draw a circle around the largest decimal

• draw a box around the smallest decimal

0.219	0.192	0.129	1.066	1.05

1.40	1.408	1.063	1.363	1.093

1.41	1.161	1.31	1.071	1.061

1.413	1.41	0.172	0.127	1.407

0.152	1.065	1.05	0.125	0.6

0.13	1.416	0.3	1.146	0.14

Look at the decimals you circled or drew a box around.
Write them on the number lines.

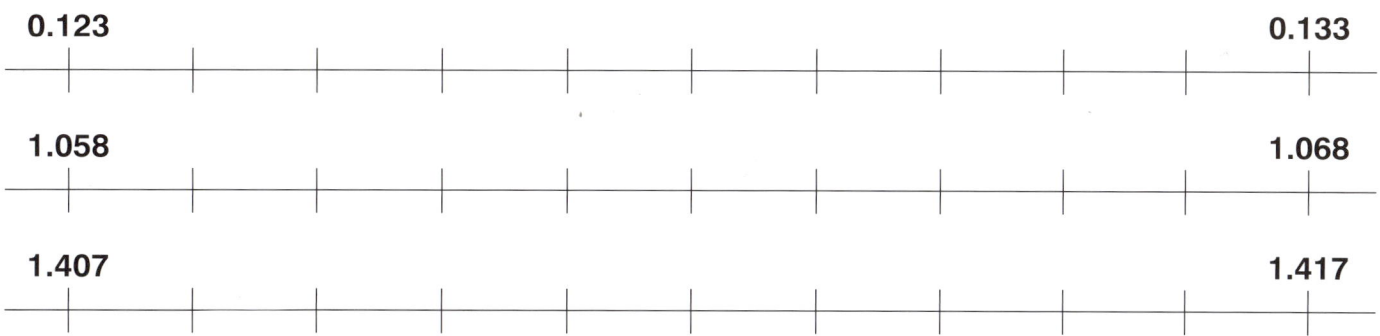

0.123   0.133

1.058   1.068

1.407   1.417

Ask your partner to check your number lines.

Talk with your partner about how you knew
where to write the decimals on the number lines.

Answers to 1A

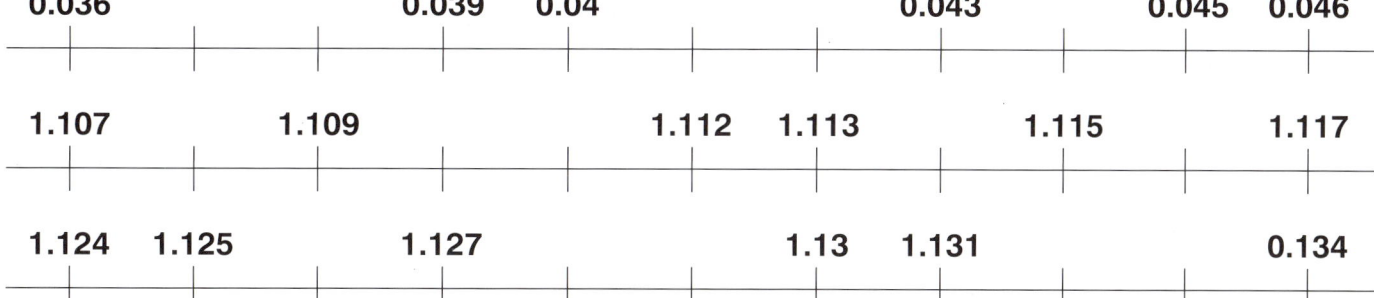

0.036   0.039   0.04   0.043   0.045   0.046

1.107   1.109   1.112   1.113   1.115   1.117

1.124   1.125   1.127   1.13   1.131   0.134

Paired activity 1A

Order decimals with up to three places, and position them on a number line [some support]

Name: _____

My partner's name: _____

Date: _____

Speak Say the decimal to your partner and write it on number line 1).

1) 0.147

Listen Listen to your partner and write the decimal they say on number line 1).

0.145

1) |___|___|___|___|___|___|___|

Speak **Listen** Do the same for all the other decimals.

2) 0.13   5) 2.3   8) 4.816
3) 1.339   6) 1.204   9) 3.018
4) 2.512   7) 5.019   10) 2.709

0.128

2) |___|___|___|___|___|___|___|

5.017

7) |___|___|___|___|___|___|___|

1.342

3) |___|___|___|___|___|___|___|

4.819

8) |___|___|___|___|___|___|___|

2.511

4) |___|___|___|___|___|___|___|

3.021

9) |___|___|___|___|___|___|___|

2.306

5) |___|___|___|___|___|___|___|

2.707

10) |___|___|___|___|___|___|___|

1.199

6) |___|___|___|___|___|___|___|

Compare your worksheet with your partner's.

Talk with your partner about any differences you notice.

Paired activity 1B

Order decimals with up to three places, and positioning them on a number line [some support]

Name: _____

My partner's name: _____

Date: _____

Listen Listen to your partner and write the decimal they say on number line 1).

0.145

1) | | | | | | | |

Speak Say this decimal to your partner and write it on number line 1).

1) 0.143

Speak **Listen** Do the same for all the other decimals.

2) 0.132   5) 2.303   8) 4.818
3) 1.34   6) 1.2   9) 3.023
4) 2.508   7) 5.02   10) 2.711

0.128

2) | | | | | | | |

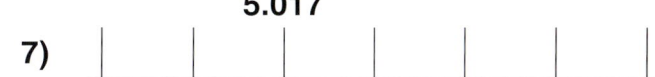

5.017

7) | | | | | | | |

1.342

3) | | | | | | | |

4.819

8) | | | | | | | |

2.511

4) | | | | | | | |

3.021

9) | | | | | | | |

2.306

5) | | | | | | | |

2.707

10) | | | | | | | |

1.199

6) | | | | | | | |

Compare your worksheet with your partner's.

Talk with your partner about any differences you notice.

Individual activity 2A

Order decimals with up to three places, and position them on a number line [less support]

Order each set of decimals, smallest to largest.

1.203	1.213	1.302	1.32	1.231					
0.134	0.128	0.138	0.126	0.14					
1.085	1.007	1.051	1.03	1.001					
2.653	2.356	2.635	2.365	2.536					
0.876	0.91	0.809	0.909	0.819					

Look at the eight decimals you have written in the grey boxes above.
Write these decimals in order, smallest to largest, in the eight grey boxes below.

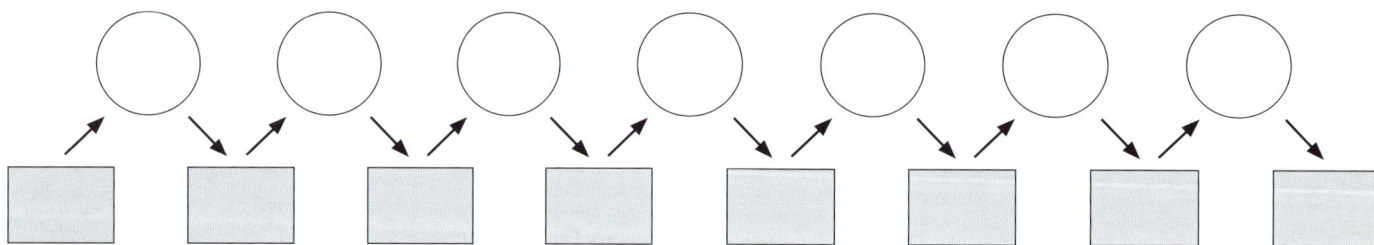

Check your answers with your partner.

Look at the decimals you have written in the grey boxes above.

Talk with your partner about what decimal you could write in each of the circles so that the decimals are still in order.

Answers to 2B

1.405	1.415	1.451	1.504	1.54
0.121	0.129	0.151	0.157	0.17
1.002	1.006	1.03	1.089	1.091
2.274	2.427	2.472	2.724	2.742
0.202	0.21	0.302	0.312	0.345

0.121	0.151	0.312	1.006	1.415	1.54	2.274	2.472

Individual activity 2B

Order decimals with up to three places, and position them on a number line [less support]

Order each set of decimals, smallest to largest.

1.405	1.451	1.504	1.415	1.54					
0.157	0.129	0.17	0.121	0.151					
1.089	1.006	1.091	1.03	1.002					
2.427	2.724	2.472	2.742	2.274					
0.345	0.21	0.302	0.202	0.312					

Look at the eight decimals you have written in the grey boxes above.

Write these decimals in order, smallest to largest, in the eight grey boxes below.

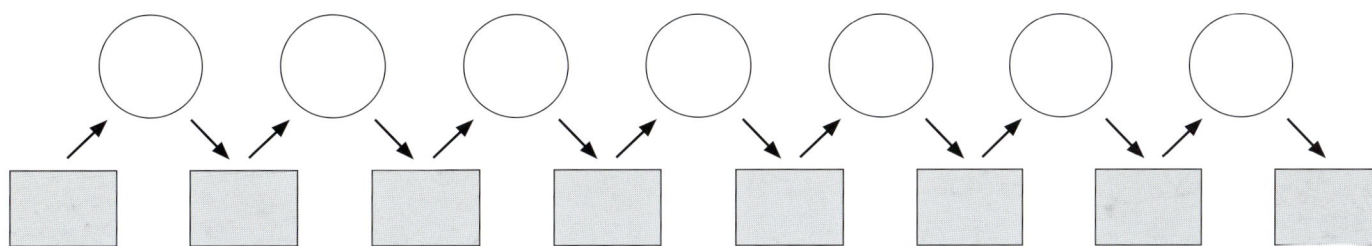

Check your answers with your partner.

Look at the decimals you have written in the grey boxes above.

Talk with your partner about what decimal you could write in each of the circles so that the decimals are still in order.

Answers to 2A

1.203	1.213	1.231	1.302	1.32
0.126	0.128	0.134	0.138	0.14
1.001	1.007	1.03	1.051	1.085
2.356	2.365	2.536	2.635	2.653
0.809	0.819	0.876	0.909	0.91

0.128	0.14	0.809	1.001	1.03	1.203	1.302	2.635

Paired activity 2A

Order decimals with up to three places, and position them on a number line [less support]

Name: _____

My partner's name: _____

Date: _____

Speak Say the decimal in the left box to your partner.

1) 0.163

Listen Listen to your partner and write the decimal they say in the empty box.

Circle the larger decimal.
Write a decimal that lies between these two decimals in the hexagon.

Speak **Listen** Do the same for all the other decimals.

2) 1.517  6) 1.7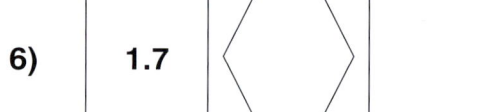

3) 0.207 7) 2.408

4) 2.001 8) 1.623

5) 0.143 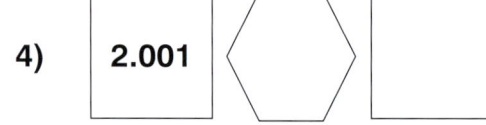 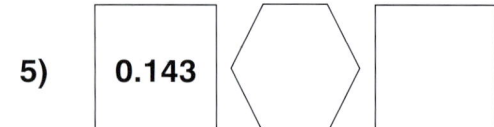 9) 1.328

Look at all the decimals you have circled.
Write these decimals in order, from smallest to largest, in the nine circles below.

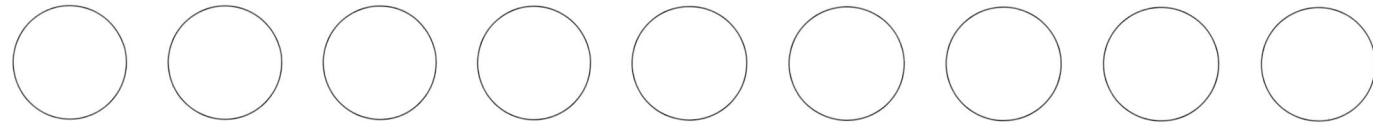

Compare your ordering of the decimals with your partner's. Are they the same?

Talk with your partner about ordering the decimals you wrote in the hexagons.

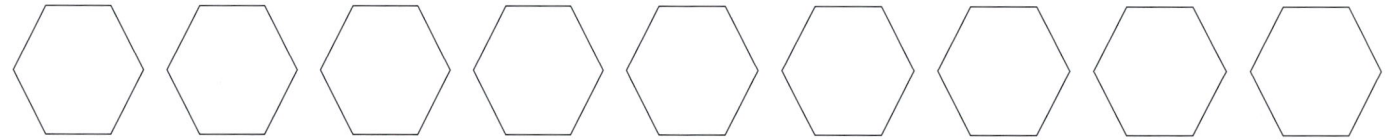

Paired activity 2B

Order decimals with up to three places, and position them on a number line [less support]

Name: _____

My partner's name: _____

Date: _____

Listen Listen to your partner and write the decimal they say in the empty box.

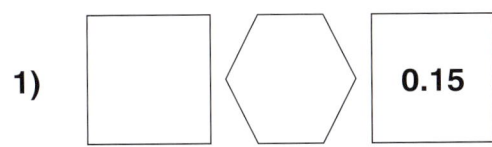

1) [ ] < > 0.15

Speak Say the decimal in the right box to your partner.

Circle the larger decimal.
Write a decimal that lies between these two decimals in the hexagon.

Speak **Listen** Do the same for all the other decimals.

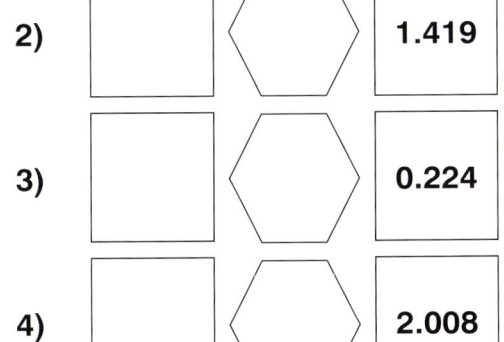
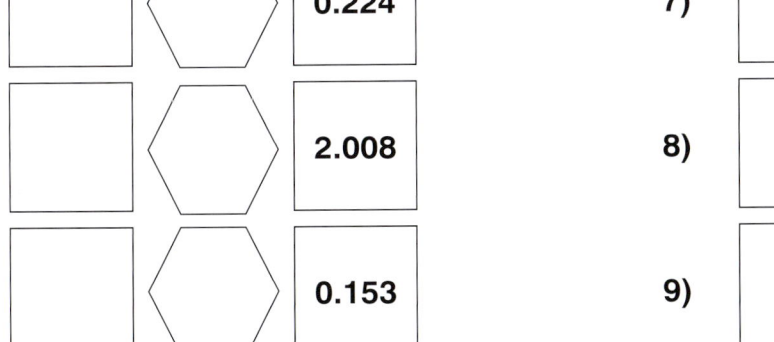

2) [ ] < > 1.419

3) [ ] < > 0.224

4) [ ] < > 2.008

5) [ ] < > 0.153

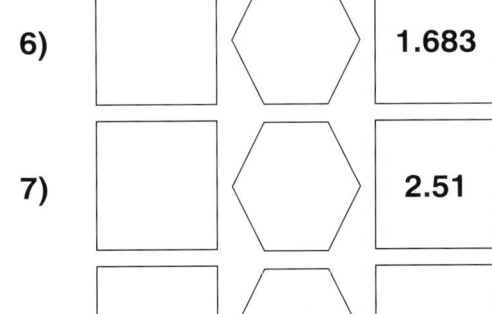
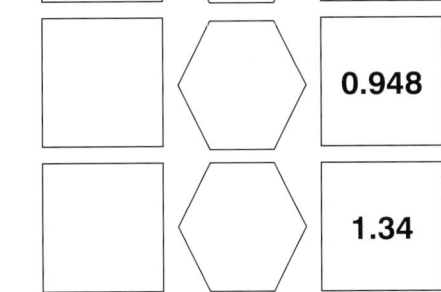

6) [ ] < > 1.683

7) [ ] < > 2.51

8) [ ] < > 0.948

9) [ ] < > 1.34

Look at all the decimals you have circled.
Write these decimals in order, from smallest to largest, in the nine circles below.

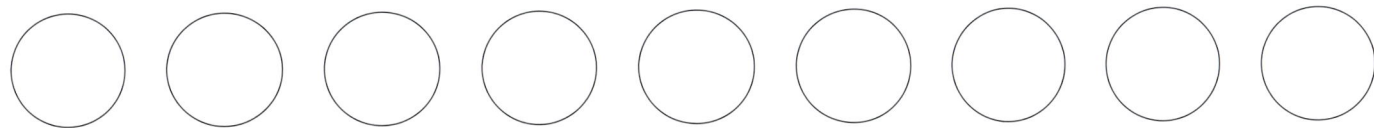

Compare your ordering of the decimals with your partner's. Are they the same?

Talk with your partner about ordering the decimals you wrote in the hexagons.

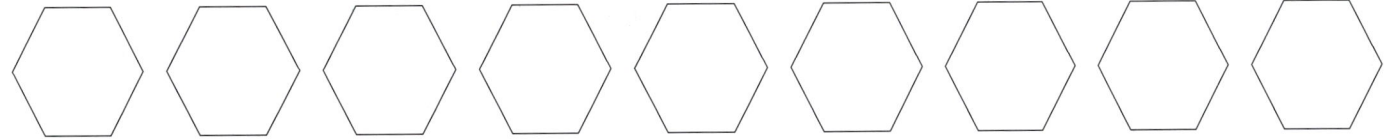

Section 2

Place value and partitioning

Level 1
- Partition decimals with three places

Level 2
- Partition decimals with three places [more complex]

Strategic approaches to develop fluency in place value and partitioning decimals with three places

 ## Have a place value chart on the wall to rehearse fluent naming of numbers

The place value chart was invented by Caleb Gattegno, a mathematics educator, and is sometimes referred to as a Gattegno chart. It looks like this:

10	20	30	40	50	60	70	80	90
1	2	3	4	5	6	7	8	9
0.1	0.2	0.3	0.4	0.5	0.6	0.7	0.8	0.9
0.01	0.02	0.03	0.04	0.05	0.06	0.07	0.08	0.09
0.001	0.002	0.003	0.004	0.005	0.006	0.007	0.008	0.009

Tapping out a number by touching a value in different rows on the chart, children can say the decimal number either individually or as a group. You can also say a decimal and invite children to come and tap out the number.

You could place this on the wall near the door so that children have to say or tap a number individually as they leave the class. Such activities need to be done very often for short bursts of time, rather than being thought of as something that you might do in only a few lessons for longer.

 ## Place value grid

It is important that children make the connection between decimal fractions and place value. Drawing on their experiences with whole numbers and using a place value grid is a powerful visual image in demonstrating this.

Remind children that the decimal point indicates that the numbers that come after it (on the right-hand side) are parts of the whole.

Just as numbers increase in size by a factor of 10 as you move up a row on the Gattegno chart, so numbers decrease by a factor of 10 as you move down a row.

Tens	Units		tenths	hundredths	thousandths
		•			

The place value grid also shows this by looking at how multiplying by 10 (or 100) moves a digit one (or two) places to the left and dividing by 10 (or 100) moves a digit one (or two) places to the right.

 ## Place value cards

Place value cards are another useful visual image in demonstrating to children how to partition decimals.

5.273 | 5 • 2 | 7 | 3 |

Expanded notation

When working with a Gattegno chart, place value grid or place value cards to demonstrate place value and partitioning, also show the children how they can write numbers using expanded notation: for example, 5.273 can be written as $5 + 0.2 + 0.07 + 0.003$.

Once children are secure with partitioning decimals into their place values, show how they can partition numbers in different ways.

$5 + 0.1 + 0.17 + 0.003$

$4 + 1.2 + 0.07 + 0.003$

$5 + 0.2 + 0.06 + 0.013$

 ## Move between representations

Children need to be fluent in moving between different representations of decimals: spoken, recorded in numerals, tapped out on the place value chart and using a place value grid or place value cards.

Regularly start with one of these representations and get the children to set up the others. Which do they find easiest to start with? Which is the most challenging?

Individual and paired activities

 ## Level 1 Partition decimals with three places

The individual activity provides practice in writing and partitioning decimals using the Gattegno chart. It also asks the children to identify decimal numbers written in expanded notation.

What number is shown on this chart?

Show me the number 9.412 on this chart.

Could you partition 1.842 in a different way?

The paired activity provides practice in asking questions to identify decimal numbers.

Which of these two numbers has the more hundreds in it? How do you know?

What were good questions to ask your partner to help you identify one of their decimal numbers?

 ## Level 2 Partition decimals with three places [more complex]

The individual activity provides expanded notation calculations involving decimals with three places. When pairs of children compare their completed sheets, they should have the same numbers written in the boxes if both children are correct.

What number is missing from this calculation? How do you know?

What does the 8 stand for in the decimal 7.852? What about the 2? … 5? … 7?

The paired activity provides further practice in partitioning decimals using expanded notation. It also involves the children saying decimals with up to three places. When pairs of children compare their completed sheets, they should each have the same set of decimals if both children are correct.

How did you partition 4.213?

What number is this? $6 + 0.9 + 0.04 + 0.002$

Further activities to develop fluency

 ## Making decimals

Provide each child with a U, t, h, th place value grid and each pair with a set of 0–9 digit cards and a coin. Children shuffle the cards and place them in a pile, face down. They take turns to take the top card from the pack and position it on one of their columns on their grid. Once a card is put on the grid, it cannot be moved.

Children continue taking cards until all the columns on both grids contain a digit card. Each child reads out the decimal number they have made. Children then toss the coin. Heads: the larger decimal wins; tails: the smaller decimal wins.

Units		tenths	hundredths	thousandths
3	•	2	5	7

Variation:

Use different place value grids.

Units		tenths	hundredths
	•		

Tens	Units		tenths	hundredths
		•		

Tens	Units		tenths	hundredths	thousandths
		•			

 ## How many decimals can you make?

Children roll a 1–9 dice four times to get four digits: for example, 1, 6, 7, 2. They arrange all four digits and a decimal point in as many different ways as they can to make decimals with up to three places: 1.672, 6.217, 6.172, 61.72, 136.7 … They read out all the decimals they make.

You may then like to ask the children to order their decimals, smallest to largest.

 ## Representing place value

Provide each pair or group with a place value grid, a set of 0–9 digit cards, a Gattegno chart, some counters, a set of decimal place value cards and paper and pencil.

Children work together to represent a decimal with up to three places, for example, 13.826, using each of the different forms of apparatus, and write it using expanded notation:

$$13.826 = 10 + 3 + 0.8 + 0.02 + 0.006$$
$$= 10 + 2 + 1.8 + 0.02 + 0.006$$
$$= 10 + 3 + 0.7 + 0.12 + 0.006$$
$$= 10 + 3 + 0.8 + 0.01 + 0.016$$

Tens	Units		tenths	hundredths	thousandths
1	3	•	8	2	6

| 1 | 3 | •8 | 2 | 6 |

10	20	30	40	50	60	70	80	90
1	2	3	4	5	6	7	8	9
0.1	0.2	0.3	0.4	0.5	0.6	0.7	0.8	0.9
0.01	0.02	0.03	0.04	0.05	0.06	0.07	0.08	0.09
0.001	0.002	0.003	0.004	0.005	0.006	0.007	0.008	0.009

My record sheet

Name: _____

Date: _____

	Before the activities			After the activities		
I can read decimal numbers correctly.	☺	😐	☹	☺	😐	☹
I can say what the value of each digit is in a decimal number.	☺	😐	☹	☺	😐	☹
I can partition a decimal with up to three places into tenths, hundredths and thousandths.	☺	😐	☹	☺	😐	☹
I can partition decimals in different ways.	☺	😐	☹	☺	😐	☹

 After the activities

These decimals all have 2 thousandths.	
These decimals all have 7 hundredths and 4 thousandths.	
I can partition the decimal 1.385 in all these different ways.	
I can partition the decimal 23.476 in all these different ways.	

Individual activity 1A

Partition decimals with three places

Write the decimals.

1)

1	2	3	4	5	6	7	8	9
0.1	0.2	0.3	0.4	0.5	0.6	0.7	0.8	0.9
0.01	0.02	0.03	0.04	0.05	0.06	0.07	0.08	0.08
0.001	0.002	0.003	0.004	0.005	0.006	0.007	0.008	0.009

2)

1	2	3	4	5	6	7	8	9
0.1	0.2	0.3	0.4	0.5	0.6	0.7	0.8	0.9
0.01	0.02	0.03	0.04	0.05	0.06	0.07	0.08	0.08
0.001	0.002	0.003	0.004	0.005	0.006	0.007	0.008	0.009

Partition each of these decimals into units, tenths, hundredths and thousandths.

3)

1	2	3	4	5	6	7	8	9
0.1	0.2	0.3	0.4	0.5	0.6	0.7	0.8	0.9
0.01	0.02	0.03	0.04	0.05	0.06	0.07	0.08	0.08
0.001	0.002	0.003	0.004	0.005	0.006	0.007	0.008	0.009

1.253

4)

1	2	3	4	5	6	7	8	9
0.1	0.2	0.3	0.4	0.5	0.6	0.7	0.8	0.9
0.01	0.02	0.03	0.04	0.05	0.06	0.07	0.08	0.08
0.001	0.002	0.003	0.004	0.005	0.006	0.007	0.008	0.009

9.412

Complete each of the following calculations.

5) $5 + 0.9 + 0.01 + 0.003 =$ ☐

6) $0.5 + 2 + 0.007 + 0.08 =$ ☐

7) $0.8 + 1 + 0.002 + 0.04 =$ ☐

8) $0.07 + 0.004 + 0.6 + 3 =$ ☐

Check your answers with your partner.

Talk with your partner about other ways you could partition the numbers.

Individual activity 1B

Partition decimals with three places

Write the decimals.

1)

1	2	3	4	5	6	7	8	9
0.1	0.2	0.3	0.4	0.5	0.6	0.7	0.8	0.9
0.01	0.02	0.03	0.04	0.05	0.06	0.07	0.08	0.08
0.001	0.002	0.003	0.004	0.005	0.006	0.007	0.008	0.009

2)

1	2	3	4	5	6	7	8	9
0.1	0.2	0.3	0.4	0.5	0.6	0.7	0.8	0.9
0.01	0.02	0.03	0.04	0.05	0.06	0.07	0.08	0.08
0.001	0.002	0.003	0.004	0.005	0.006	0.007	0.008	0.009

Partition each of these decimals into units, tenths, hundredths and thousandths.

3)

1	2	3	4	5	6	7	8	9
0.1	0.2	0.3	0.4	0.5	0.6	0.7	0.8	0.9
0.01	0.02	0.03	0.04	0.05	0.06	0.07	0.08	0.08
0.001	0.002	0.003	0.004	0.005	0.006	0.007	0.008	0.009

1.237

4)

1	2	3	4	5	6	7	8	9
0.1	0.2	0.3	0.4	0.5	0.6	0.7	0.8	0.9
0.01	0.02	0.03	0.04	0.05	0.06	0.07	0.08	0.08
0.001	0.002	0.003	0.004	0.005	0.006	0.007	0.008	0.009

5.695

Complete each of the following calculations.

5) $2 + 0.8 + 0.07 + 0.001 =$

7) $0.04 + 0.009 + 0.3 + 1 =$

6) $0.5 + 0.01 + 3 + 0.002 =$

8) $0.008 + 2 + 0.03 + 0.9 =$

Check your answers with your partner.

Talk with your partner about other ways you could partition the numbers.

Answers to 1A

1) | 4.725 | 2) | 3.961 | 3) and 4) Check your partner's answers. Do you agree?

5) | 5.913 | 6) | 2.587 | 7) | 1.842 | 8) | 3.674 |

Paired activity 1A

Partition decimals with three places

Name:

My partner's name:

Date:

You need:
- coloured pencil

Colour 8 of these decimals.
Don't let your partner see the 8 decimals you have coloured.
Your partner will also colour 8 decimals.

0.102	2.624	0.245	2.705	1.268
1.316	0.132	2.801	0.321	1.103
0.342	1.536	0.423	1.308	2.812
0.231	2.054	1.416	0.405	0.532
2.413	1.461	0.113	2.067	1.038

Speak Take turns to ask questions similar to these to identify your partner's 8 decimals.

- Do any of your decimals have a 3 in the hundredths place?
- Do any of your decimals have 2 units?
- Do any of your decimals have a 5 in them?
- Does one of your decimals have 4 thousandths?

You and your partner have the same numbers in your grid. Use the numbers in this grid to help you identify your partner's 8 decimals.

As you ask your questions, cross out decimals you know your partner hasn't coloured.

When you identify one of your partner's 8 decimals, draw a circle round that decimal.

The winner is the first player to identify their partner's 8 decimals.

Talk with your partner about what were good questions to ask.

0.102	2.624	0.245	2.705	1.268
1.316	0.132	2.801	0.321	1.103
0.342	1.536	0.423	1.308	2.812
0.231	2.054	1.416	0.405	0.532
2.413	1.461	0.113	2.067	1.038

Paired activity 1B

Partition decimals with three places

Name: _____

My partner's name: _____

Date: _____

You need:
- coloured pencil

Colour 8 of these decimals.
Don't let your partner see the 8 decimals you have coloured.
Your partner will also colour 8 decimals.

0.102	2.624	0.245	2.705	1.268
1.316	0.132	2.801	0.321	1.103
0.342	1.536	0.423	1.308	2.812
0.231	2.054	1.416	0.405	0.532
2.413	1.461	0.113	2.067	1.038

Speak Take turns to ask questions similar to these to identify your partner's 8 decimals.

- Do any of your decimals have a 3 in the hundredths place?
- Do any of your decimals have 2 units?
- Do any of your decimals have a 5 in them?
- Does one of your decimals have 4 thousandths?

You and your partner have the same numbers in your grid. Use the numbers in this grid to help you identify your partner's 8 decimals.

As you ask your questions, cross out decimals you know your partner hasn't coloured.

When you identify one of your partner's 8 decimals, draw a circle round that decimal.

The winner is the first player to identify their partner's 8 decimals.

Talk with your partner about what were good questions to ask.

0.102	2.624	0.245	2.705	1.268
1.316	0.132	2.801	0.321	1.103
0.342	1.536	0.423	1.308	2.812
0.231	2.054	1.416	0.405	0.532
2.413	1.461	0.113	2.067	1.038

Individual activity 2A

Partition decimals with three places [more complex]

Complete each of the following calculations.

1) $2.645 = 2 + \boxed{} + 0.04 + 0.005$

7) $6.879 = 6 + 0.07 + 0.8 + \boxed{}$

2) $5.387 = \boxed{} + 0.3 + 0.08 + 0.007$

8) $2.963 = 2 + 0.9 + 0.003 + \boxed{}$

3) $3.154 = 3 + 0.1 + 0.05 + \boxed{}$

9) $9.341 = 0.3 + 0.04 + \boxed{} + 0.001$

4) $1.741 = 1 + 0.7 + \boxed{} + 0.001$

10) $8.753 = \boxed{} + 0.7 + 0.05 + 8$

5) $2.976 = 2 + 0.9 + \boxed{} + 0.006$

11) $4.376 = 0.3 + \boxed{} + 4 + 0.006$

6) $7.852 = 7 + \boxed{} + 0.05 + 0.002$

12) $5.921 = 0.02 + \boxed{} + 5 + 0.001$

Compare your answers with your partner's answers. What do you notice?

Talk with your partner about any answers that are different.

Individual activity 2B

Partition decimals with three places [more complex]

Complete each of the following calculations.

1) $1.682 = 1 + \boxed{} + 0.08 + 0.002$

2) $5.714 = \boxed{} + 0.7 + 0.01 + 0.004$

3) $2.374 = 2 + 0.3 + 0.07 + \boxed{}$

4) $8.246 = 8 + 0.2 + \boxed{} + 0.006$

5) $4.175 = 4 + 0.1 + \boxed{} + 0.005$

6) $1.834 = 1 + \boxed{} + 0.03 + 0.004$

7) $3.249 = 3 + 0.04 + 0.2 + \boxed{}$

8) $7.269 = 7 + 0.2 + 0.009 + \boxed{}$

9) $9.185 = 0.1 + 0.08 + \boxed{} + 0.005$

10) $3.643 = \boxed{} + 0.6 + 0.04 + 3$

11) $6.273 = 0.2 + \boxed{} + 6 + 0.003$

12) $2.973 = 0.07 + \boxed{} + 2 + 0.003$

Compare your answers with your partner's answers.
What do you notice?

Talk with your partner about any answers that are different.

Paired activity 2A

Name:

My partner's name:

Date:

Partition decimals with three places [more complex]

Partition each of these decimals.

For example: (6.347) = | 6 | + | 0.3 | + | 0.04 | + | 0.007 |

1) (2.852) = [ ] + [ ] + [ ] + [ ]

2) (5.798) = [ ] + [ ] + [ ] + [ ]

3) (1.374) = [ ] + [ ] + [ ] + [ ]

4) (4.213) = [ ] + [ ] + [ ] + [ ]

5) (3.916) = [ ] + [ ] + [ ] + [ ]

Speak Say each of the calculations you partitioned above to your partner, but not the decimals in the ovals.

Listen Listen to your partner and write the calculations in the boxes below.
Don't work out the answers yet.

1) ( ) = [ ] + [ ] + [ ] + [ ]

2) ( ) = [ ] + [ ] + [ ] + [ ]

3) ( ) = [ ] + [ ] + [ ] + [ ]

4) ( ) = [ ] + [ ] + [ ] + [ ]

5) ( ) = [ ] + [ ] + [ ] + [ ]

Now work out the answer to each of the calculations above, writing the answer in the oval.

Check your answers with your partner.

Talk with your partner about any of the answers that are different.

Paired activity 2B

**Partition decimals with three places
[more complex]**

Partition each of these decimals.

For example: (6.347) = [6] + [0.3] + [0.04] + [0.007]

1) (1.684) = [   ] + [   ] + [   ] + [   ]

2) (5.813) = [   ] + [   ] + [   ] + [   ]

3) (4.769) = [   ] + [   ] + [   ] + [   ]

4) (6.942) = [   ] + [   ] + [   ] + [   ]

5) (2.351) = [   ] + [   ] + [   ] + [   ]

Listen Listen to your partner and write the calculations
in the boxes below.
Don't work out the answers yet.

1) (   ) = [   ] + [   ] + [   ] + [   ]

2) (   ) = [   ] + [   ] + [   ] + [   ]

3) (   ) = [   ] + [   ] + [   ] + [   ]

4) (   ) = [   ] + [   ] + [   ] + [   ]

5) (   ) = [   ] + [   ] + [   ] + [   ]

Speak Say each of the calculations you partitioned at the top of the
sheet to your partner, but not the decimals in the ovals.

Now work out the answer to each of the calculations above,
writing the answer in the oval.

Check your answers with your partner.

Talk with your partner about any of the answers that are different.

Understanding and using tens

Level 1

- Use knowledge of place value and multiplication facts to 10 × 10 to derive multiplication and division facts involving decimals U.t × U, U.t ÷ U [restricted range of numbers]

Level 2

- Use knowledge of place value and multiplication facts to 10 × 10 to derive multiplication and division facts involving decimals U.t × U, U.t ÷ U [wider range of numbers]

Strategic approaches to develop fluency in multiplication and division facts involving decimals

 ## Use knowledge of multiplication and division facts and place value

Make sure that children can confidentlymultiply and divide by 10 and 100 and that they understand that multiplying by 10 gives an answer that is 10 times bigger than the original number and all the digits move one place to the left, while dividing by 10 gives an answer that is 10 times smaller than the original number and all the digits move one place to the right.

Start with known multiplication facts before relating these to decimal multiplication facts: for example, count on and back in steps of 4 before relating this to counting on and back in steps of 0.4. Encourage children to explain the relationship between the two sets of numbers.

 ## Use number lines and the notion of scale factors

Show children the multiples of 4 and the related multiples involving decimals on scale factor number lines and discuss the similarities.

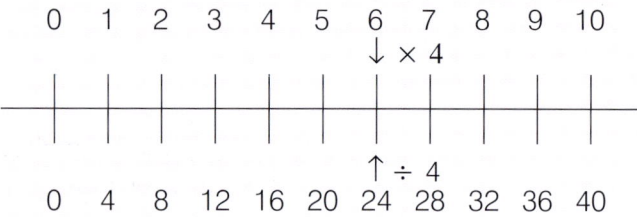 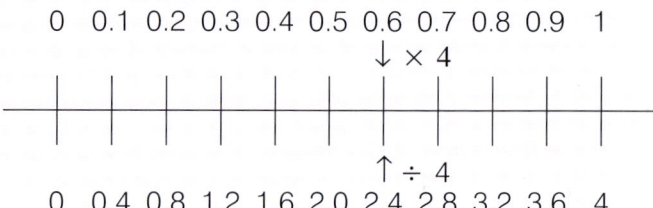

 ## Focus on multiplication

In a sense, no one ever calculates a division. When asked to calculate, for example, $35 \div 5$, fluent calculators know that $7 \times 5 = 35$ and use their knowledge of that multiplication fact to know quickly that $35 \div 5 = 7$.

Encourage children to talk about related multiplication and division facts. For example, there are four related facts using the trio 5, 7, 35: two multiplication and two division.

$$5 \times 7 = 35$$

$$7 \times 5 = 35$$

$$35 \div 5 = 7$$

$$35 \div 7 = 5$$

By firmly embedding the trio 5, 7, 35, all the related multiplication and division facts become fluent.

In the same way, reinforce the division facts corresponding to multiplication facts involving decimals.

$$0.5 \times 7 = 3.5$$

$$0.7 \times 5 = 3.5$$

$$3.5 \div 5 = 0.7$$

$$3.5 \div 7 = 0.5$$

 ## How to express calculations in different ways

Make sure that children meet and can interpret multiplication and division calculations that are written in a variety of different ways.

$$\square \times 5 = 3.5 \qquad 0.7 \times \square = 3.5$$

$$3.5 = 0.7 \times \square \qquad 3.5 = \square \times 5$$

$$\square \div 5 = 0.7 \qquad 3.5 \div \square = 0.7$$

$$0.7 = 3.5 \div \square \qquad 0.7 = \square \div 5$$

Individual and paired activities

Level 1 Use knowledge of place value and multiplication facts to
10×10 to derive multiplication and division facts involving decimals
U.t $\times$ U, U.t $\div$ U [restricted range of numbers]

The individual activity provides practice in the multiplication and related division facts to 10×10, and using and applying these to derive the answers to similar calculations involving decimals.

What is 24 divided by 3? If you know that 24 divided by 3 is 8, then what is 2.4 divided by 3?

If you know that 9 times 3 is 27, what other multiplication number sentences do you know involving decimals? What division number sentences do you know that are related to 9 times 3? What about some involving decimals?

The paired activity provides further practice in using knowledge of place value and multiplication facts to 10×10 to derive multiplication and division facts involving decimals (U.t x U, U.t $\div$ U) and also an opportunity for children to discuss their answers and mental methods with their partner.

If 3 multiplied by 6 is 18, then what is 0.3×6? Why?

What fact did you use to work out the answer to 2.4 divided by 4?

Level 2 Use knowledge of place value and multiplication facts to
10×10 to derive multiplication and division facts involving decimals
U.t $\times$ U, U.t $\div$ U [wider range of numbers]

In the individual activity, children mark a set of 20 multiplication and division calculations involving decimals (U.t $\times$ U, U.t $\div$ U). When each child has completed marking their sheet, they check with their partner which answers were correct and which were incorrect. They then talk with their partner about each of the incorrect answers and write the correct answer beside each calculation.

Does 5.6 divided by 8 equal 0.6? What should the answer be?

In the paired activity, children independently answer a set of calculations involving multiplication and related division facts to 10×10. They then listen and calculate as their partner reads out related facts involving decimals (U.t $\times$ U, U.t $\div$ U).

What is the answer to 0.9 times 4?

Which fact did you use to help you work out the answer to this?

Further activities to develop fluency

Multiplication trios

Prepare a set of demonstration cards that show multiplication trios.

Hold up a card and hide one of the numbers. Children suggest different multiplication and division calculations, using the two visible numbers.

3.5 divided by 5 equals?

5 times what makes 3.5?

How many times does 5 divide into 3.5?

What do you need to multiply 5 by to get 3.5?

What times table fact did you use to help you work out the answer?

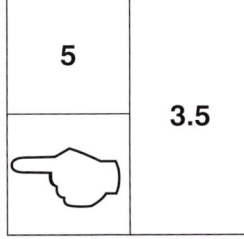

Multiplication squares

Display a multiplication square for the tables to 10×10 and discuss with the children its features and how to use it, including how to find the related division facts.

Make sure that the children are familiar with this before displaying a multiplication square for the related facts involving decimals.

Discuss the following with the children:

- Just as $4 \times 7 = 28$, so too does $0.4 \times 7 = 2.8$.
- You can do multiplication in any order: $0.4 \times 7 = 2.8$ and $7 \times 0.4 = 2.8$
- Multiplication is the inverse of division: if you know that $0.4 \times 7 = 2.8$, you also know that $2.8 \div 7 = 0.4$.

×	1	2	3	4	5	6	7	8	9	10
1	1	2	3	4	5	6	7	8	9	10
2	2	4	6	8	10	12	14	16	18	20
3	3	6	9	12	15	18	21	24	27	30
4	4	8	12	16	20	24	28	32	36	40
5	5	10	15	20	25	30	35	40	45	50
6	6	12	18	24	30	36	42	48	54	60
7	7	14	21	28	35	42	49	56	63	70
8	8	16	24	32	40	48	56	64	72	80
9	9	18	27	36	45	54	63	72	81	90
10	10	20	30	40	50	60	70	80	90	100

×	0.1	0.2	0.3	0.4	0.5	0.6	0.7	0.8	0.9	1
0.1	0.1	0.2	0.3	0.4	0.5	0.6	0.7	0.8	0.9	1
0.2	0.2	0.4	0.6	0.8	1	1.2	1.4	1.6	1.8	2
0.3	0.3	0.6	0.9	1.2	1.5	1.8	2.1	2.4	2.7	3
0.4	0.4	0.8	1.2	1.6	2	2.4	2.8	3.2	3.6	4
0.5	0.5	1	1.5	2	2.5	3	3.5	4	4.5	5
0.6	0.6	1.2	1.8	2.4	3	3.6	4.2	4.8	5.4	6
0.7	0.7	1.4	2.1	2.8	3.5	4.2	4.9	5.6	6.3	7
0.8	0.8	1.6	2.4	3.2	4	4.8	5.6	6.4	7.2	8
0.9	0.9	1.8	2.7	3.6	4.5	5.4	6.3	7.2	8.1	9
1	1	2	3	4	5	6	7	8	9	10

 ## Point to it – decimals

Write the 10 multiples of 0.4 from 0.4 to 4 in random order near each other on the board.

```
            3.2    2
0.4    2.8
               4    1.6
0.8    1.2
            2.4    3.6
```

Give two children a ruler each and ask them to stand either side of the numbers. Ask a multiplication fact where the answer is on the board: for example, "What is 0.4 times 6?" Children point to the answer: 2.4.

The first child to point to the answer stays in, the other child sits down.

Another child comes out to the board. Continue as above. Which child can stay in the longest?

Variations:
Ask all the children in the class to think of a calculation where the answer is on the board. The rest of the class asks the two children at the board their calculations.

Write the 10 tenths from 0.1 to 1 on the board and ask a division fact where the answer is on the board: for example, "What is 2.4 divided by 4?"

My record sheet

	Before the activities	After the activities
I can use place value and my tables to work out multiplication facts for decimals, such as 0.8×7.	☺ 😐 ☹	☺ 😐 ☹
I can use place value and my tables to work out division facts for decimals, such as $4.8 \div 6$.	☺ 😐 ☹	☺ 😐 ☹

 ## After the activities

Here are some multiplication facts involving decimals.	
Here are some division facts involving decimals.	
By using the multiplication fact $7 \times 4 = 28$, I also know these related decimal facts.	
By using the division fact $24 \div 3 = 8$, I also know these related decimal facts.	

Individual activity 1A

Name: _____

Date: _____

Derive multiplication and division facts involving decimals U.t × U, U.t ÷ U [restricted range of numbers]

Complete each of the following sets of calculations.

1)
$5 \times 8 =$

$0.5 \times 8 =$

2)
$7 \times 4 =$

$0.7 \times 4 =$

3)
$9 \times 3 =$

$0.9 \times 3 =$

4)
$4 \times 6 =$

$0.4 \times 6 =$

5)
$16 \div 2 =$

$1.6 \div 2 =$

6)
$2 \times 2 =$

$0.2 \times 2 =$

7)
$42 \div 6 =$

$4.2 \div 6 =$

8)
$24 \div 3 =$

$2.4 \div 3 =$

9)
$36 \div 4 =$

$3.6 \div 4 =$

10)
$3 \times 5 =$

$0.3 \times 5 =$

11)
$12 \div 6 =$

$1.2 \div 6 =$

12)
$35 \div 5 =$

$3.5 \div 5 =$

Check your answers with your partner.

Talk with your partner about any patterns you notice.

Answers to 1B

1)
18
1.8

2)
20
2

3)
3
0.3

4)
45
4.5

5)
8
0.8

6)
9
0.9

7)
7
0.7

8)
27
2.7

9)
36
3.6

10)
6
0.6

11)
9
0.9

12)
32
3.2

Individual activity 1B

Name:

Date:

Derive multiplication and division facts involving decimals U.t × U, U.t ÷ U [restricted range of numbers]

Complete each of the following sets of calculations.

1) $3 \times 6 =$

$0.3 \times 6 =$

2) $5 \times 4 =$

$0.5 \times 4 =$

3) $9 \div 3 =$

$0.9 \div 3 =$

4) $5 \times 9 =$

$0.5 \times 9 =$

5) $48 \div 6 =$

$4.8 \div 6 =$

6) $18 \div 2 =$

$1.8 \div 2 =$

7) $21 \div 3 =$

$2.1 \div 3 =$

8) $9 \times 3 =$

$0.9 \times 3 =$

9) $6 \times 6 =$

$0.6 \times 6 =$

10) $24 \div 4 =$

$2.4 \div 4 =$

11) $45 \div 5 =$

$4.5 \div 5 =$

12) $8 \times 4 =$

$0.8 \times 4 =$

Check your answers with your partner.

Talk with your partner about any patterns you notice.

Answers to 1A

1) 40 / 4

2) 28 / 2.8

3) 27 / 2.7

4) 24 / 2.4

5) 8 / 0.8

6) 4 / 0.4

7) 7 / 0.7

8) 8 / 0.8

9) 9 / 0.9

10) 15 / 1.5

11) 2 / 0.2

12) 7 / 0.7

Paired activity 1A

Name:

My partner's name:

Date:

Derive multiplication and division facts involving decimals U.t × U, U.t ÷ U [restricted range of numbers]

Answer the calculations in the white boxes.
Then draw lines to match each white box to the related grey box.
Now answer the calculations in the grey boxes.
One pair of boxes has been done for you.

$3 \times 2 = 6$

$5 \times 5 =$

$0.4 \times 3 =$

$0.5 \times 5 =$

$0.8 \times 2 =$

$1.2 \div 2 =$

$45 \div 5 =$

$27 \div 3 =$

$18 \div 6 =$

$2.7 \div 3 =$

$1.8 \div 6 =$

$0.3 \times 2 = 0.6$

$8 \times 2 =$

$6 \times 4 =$

$4 \times 3 =$

$0.6 \times 4 =$

$28 \div 4 =$

$2.8 \div 4 =$

$4.5 \div 5 =$

$12 \div 2 =$

Speak Say one of the facts in the white boxes to your partner, but not the answer.

Now say the related fact in the grey box to your partner, but not the answer.
Do you both agree on the answers? If not, talk about it.

Listen Listen to the two calculations your partner asks you and work out the answers.

Tell your partner the answers.
Do you both agree on the answers? If not, talk about it.

Continue until you have asked each other all 10 pairs of calculations.

Paired activity 1B

Name: _____

My partner's name: _____

Date: _____

Derive multiplication and division facts
involving decimals U.t × U, U.t ÷ U
[restricted range of numbers]

Answer the calculations in the white boxes.
Then draw lines to match each white box to the related grey box.
Now answer the calculations in the grey boxes.
One pair of boxes has been done for you.

$3 \times 8 = 24$		$16 \div 2 =$
	$1.6 \div 2 =$	
$0.5 \times 6 =$		$4.2 \div 6 =$
	$0.3 \times 6 =$	
$2 \times 4 =$		$7 \times 2 =$
	$42 \div 6 =$	
$2.5 \div 5 =$		$2.4 \div 4 =$
	$0.3 \times 8 = 2.4$	
$5 \times 6 =$		$25 \div 5 =$
	$27 \div 3 =$	
$0.2 \times 4 =$		$3 \times 6 =$
	$0.7 \times 2 =$	
$2.7 \div 3 =$		$24 \div 4 =$

Listen · Listen to the two calculations your partner asks you
and work out the answers.

Tell your partner the answers.
Do you both agree on the answers? If not, talk about it.

Speak Say one of the facts in the white boxes to your partner, but not the answer.

Now say the related fact in the grey box to your partner, but not the answer.
Do you both agree on the answers? If not, talk about it.

Continue until you have asked each other all 10 pairs of calculations.

Individual activity 2A

Name:

Date:

Derive multiplication and division facts involving decimals U.t × U, U.t ÷ U [wider range of numbers]

Check the answer to each calculation.
Put a tick (✓) beside correct answers and a cross (✗) beside wrong answers.

1) $0.6 \times 7 = 4.2$

2) $0.8 \times 3 = 2.6$

3) $2.8 \div 7 = 0.4$

4) $6.3 \div 8 = 0.8$

5) $0.2 \times 7 = 1.2$

6) $3.6 \div 9 = 0.4$

7) $1.6 \div 4 = 0.2$

8) $0.7 \times 3 = 2.1$

9) $2.4 \div 3 = 0.9$

10) $0.4 \times 6 = 2.4$

11) $4.2 \div 6 = 0.7$

12) $0.5 \times 8 = 4.0$

13) $0.6 \times 9 = 5.4$

14) $0.3 \times 5 = 1.5$

15) $1.8 \div 2 = 0.9$

16) $2.5 \div 5 = 1.5$

17) $0.9 \times 4 = 3.6$

18) $6.4 \div 8 = 0.8$

19) $3.5 \div 7 = 0.8$

20) $0.7 \times 8 = 5.4$

Check with your partner which answers were correct and which were wrong.

Talk with your partner about each of the wrong answers.

Write the correct answer beside the calculation.

Answers to 2B

1) ✓	6) ✗	11) ✓	16) ✓
2) ✓	7) ✓	12) ✗	17) ✓
3) ✗	8) ✓	13) ✗	18) ✓
4) ✓	9) ✗	14) ✗	19) ✗
5) ✓	10) ✓	15) ✗	20) ✗

Individual activity 2B

Derive multiplication and division facts involving decimals U.t × U, U.t ÷ U [wider range of numbers]

Check the answer to each calculation.
Put a tick (✓) beside correct answers and a cross (×) beside wrong answers.

1) $0.2 \times 6 = 1.2$

2) $1.6 \div 4 = 0.4$

3) $4.2 \div 6 = 0.8$

4) $0.6 \times 8 = 4.8$

5) $2.1 \div 7 = 0.3$

6) $0.9 \times 3 = 2.8$

7) $0.8 \times 4 = 3.2$

8) $4.5 \div 5 = 0.9$

9) $5.6 \div 8 = 0.6$

10) $0.6 \times 3 = 1.8$

11) $0.7 \times 7 = 4.9$

12) $0.4 \times 9 = 3.2$

13) $5.4 \div 9 = 0.7$

14) $2.4 \div 3 = 0.7$

15) $0.9 \times 5 = 4$

16) $0.8 \div 2 = 0.4$

17) $3.2 \div 4 = 0.8$

18) $0.5 \times 6 = 3$

19) $0.4 \times 7 = 2.1$

20) $3.6 \div 6 = 0.7$

Check with your partner which answers were correct and which were wrong.

Talk with your partner about each of the wrong answers.

Write the correct answer beside the calculation.

Answers to 2A

1) ✓

2) ×

3) ✓

4) ×

5) ×

6) ✓

7) ×

8) ✓

9) ×

10) ✓

11) ✓

12) ✓

13) ✓

14) ✓

15) ✓

16) ×

17) ✓

18) ✓

19) ×

20) ×

Paired activity 2A

Name:

My partner's name:

Date:

Derive multiplication and division facts involving decimals U.t × U, U.t ÷ U [wider range of numbers]

Write the answer to each calculation in the box ☐.

1) $7 \times 4 =$ ☐ ◯

2) $36 \div 4 =$ ☐ ◯

3) $24 \div 3 =$ ☐ ◯

4) $9 \times 8 =$ ☐ ◯

5) $6 \times 7 =$ ☐ ◯

6) $49 \div 7 =$ ☐ ◯

7) $35 \div 5 =$ ☐ ◯

8) $6 \times 9 =$ ☐ ◯

9) $4 \times 3 =$ ☐ ◯

10) $48 \div 8 =$ ☐ ◯

Speak Say these calculations to your partner, but not the answers in the circles ◯.

$2.8 \div 7 = (0.4)$

$0.7 \times 9 = (6.3)$

$2.4 \div 3 = (0.8)$

$0.8 \times 9 = (7.2)$

$0.2 \times 8 = (1.6)$

$4.5 \div 5 = (0.9)$

$5.6 \div 7 = (0.8)$

$0.9 \times 4 = (3.6)$

$3.6 \div 6 = (0.6)$

$0.8 \times 6 = (4.8)$

Listen Listen to the calculations your partner tells you.

For each calculation, find the related fact at the top of the sheet and write the answer to the calculation your partner tells you in the circle ◯ next to it.

Check your answers with your partner.

Talk with your partner about how you used each number fact to work out the answer to the related fact.

Paired activity 2B

Derive multiplication and division facts involving decimals U.t × U, U.t ÷ U [wider range of numbers]

Name:

My partner's name:

Date:

Write the answer to each calculation in the box ☐.

1) $9 \times 4 =$

2) $28 \div 7 =$

3) $36 \div 6 =$

4) $2 \times 8 =$

5) $8 \times 9 =$

6) $24 \div 3 =$

7) $45 \div 5 =$

8) $7 \times 9 =$

9) $8 \times 6 =$

10) $56 \div 7 =$

Listen Listen to the calculations your partner tells you.

For each calculation, find the related fact above and write the answer to the calculation your partner tells you in the circle ◯ next to it.

Speak Say these calculations to your partner, but not the answers in the circles ◯.

$0.4 \times 3 = (1.2)$

$0.9 \times 8 = (7.2)$

$0.6 \times 9 = (5.4)$

$0.6 \times 7 = (4.2)$

$4.8 \div 8 = (0.6)$

$2.4 \div 3 = (0.8)$

$0.7 \times 4 = (2.8)$

$3.5 \times 5 = (0.7)$

$4.9 \div 7 = (0.7)$

$3.6 \div 4 = (0.9)$

Check your answers with your partner.

Talk with your partner about how you used each number fact to work out the answer to the related fact.

Section 4

Deriving and recalling addition and subtraction facts, and using that knowledge

Level 1

- Find the difference between a positive and a negative, or two negative integers, in context [restricted range of numbers]

Level 2

- Find the difference between a positive and a negative, or two negative integers, in context [wider range of numbers]

Strategic approaches to develop fluency in finding the difference between a positive and a negative, or two negative integers

 Use negative numbers in context

Use number lines, counting sticks and thermometers to show how numbers extend beyond 0 when counting back.

Encourage children to identify and discuss what the unmarked numbers are.

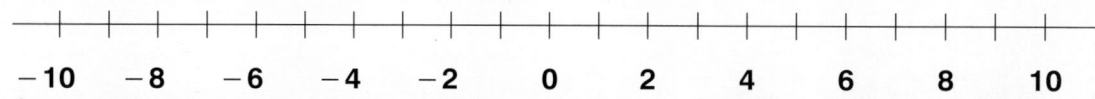

Where would you place –2 on these number lines?

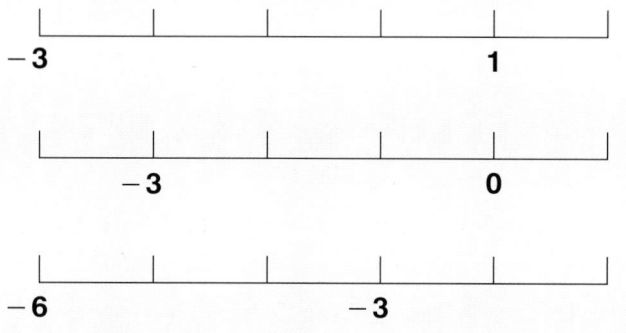

Point to part of the counting stick and ask questions such as:

If the middle number is zero, can you place –4 on the counting stick?

What is the number at the far left if the number at the far right is 8?

What happens to the number at the far left if I change the number at the far right to 5? To –5? To 0?

In the context of temperature, ask the children to:

- look in holiday brochures for temperatures from around the world and then represent these on a number line;

- find the temperatures in different cities, from newspapers or the internet, on the same day and compare and order these temperatures.

What do each of these divisions represent on this thermometer?

What does the temperature on this thermometer read?

What are the four missing temperatures from the thermometer?

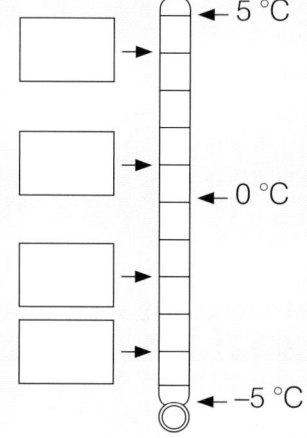

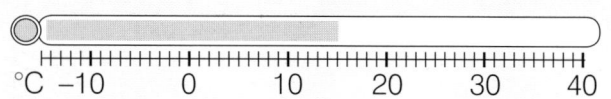

 Compare and order positive and negative numbers

Use number lines, counting sticks and thermometers to demonstrate that, for example, –5 is less than –1. Children need to be able to confidently compare and order positive and

negative integers before finding the difference between a positive and a negative, or two negative integers.

Subtraction as finding a 'difference' by counting up

Focus on subtraction involving a positive and a negative, or two negative integers, as finding the 'difference' by counting up from the smaller to the larger number. If children are confident in this concept and can use it for the subtraction of positive integers, they should be able to apply the same principles to finding the difference between a positive and a negative, or two negative integers. Use number lines, counting sticks and thermometers to assist children in their calculations.

The difference between -3 and 4

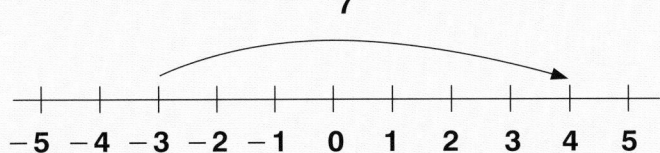

The difference between -1 and -5

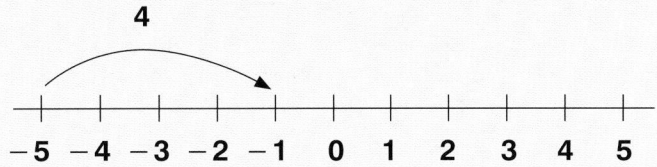

Individual and paired activities

 ## Level 1 Find the difference between a positive and a negative or two negative integers in context [restricted range of numbers]

In the individual activity, children read thermometers for different cities and use these as number lines to help them find the difference between a positive and a negative and two negative integers.

What is the temperature for Berlin? What about Moscow? Which city has the colder temperature?

What is the difference in temperatures between Berlin and Moscow?

In the paired activity, children need to share the information they have on their thermometers to find the differences in temperatures between cities.

Answers to Paired activity 1A

1) 15 °C   2) 10 °C   3) 12 °C   4) 13 °C
5) 7 °C   6) 7 °C   7) 3 °C   8) 1 °C

Answers to Paired activity 1B

1) 11 °C   2) 9 °C   3) 12 °C   4) 15 °C
5) 2 °C   6) 2 °C   7) 1 °C   8) 1 °C

Which of these cities is the coldest/hottest? Order the cities, coldest to hottest.

Which two cities have the greatest difference in temperature? How do you know that?

 ## Level 2 Find the difference between a positive and a negative or two negative integers in context [wider range of numbers]

In the individual activity, children use the weather charts to find the differences in temperatures between different cities. The temperatures in this activity have a wider range of numbers than in the Level 1 individual activity.

What are the temperatures in Singapore and Gorky? What is the difference in their temperatures?

What about for Gorky and Vancouver?

In the paired activity, children need to share the information they have about the summer and winter temperatures for different European capital cities to find the differences in temperatures between cities.

Answers to Paired activity 2A

1) 26 °C   2) 33 °C   3) 37 °C   4) 27 °C
5) 8 °C   6) 6 °C   7) 15 °C   8) 7 °C

Answers to Paired activity 2B

1) 26 °C   2) 31 °C   3) 39 °C   4) 30 °C
5) 1 °C   6) 3 °C   7) 5 °C   8) 13 °C

Which city has the coldest winter temperature? Which city has the warmest winter temperature? What is the difference in temperature between these two cities in winter?

Which city has the greatest difference between its summer and winter temperatures?

Further activities to develop fluency

 ## The difference game

Provide each pair with an addition and subtraction symbol dice (+ + + − − −), a 0–9 dice and a coin. Provide each child with pencil and paper. Children take turns to roll both dice and write down the symbol and the number. Each child rolls the two dice again and does the same, then finds the difference between the two numbers.

Children then toss the coin: heads – the larger answer wins, tails – the smaller answer wins.

Children play 10 rounds.

 ## Make a difference of 6

Children write down pairs of numbers with a difference of 6. One of the numbers must be a positive number, the other a negative number.

Variations:

Both numbers must be negative numbers.

Make a difference of 5/10/11 …

 ## Odd or even difference

Provide each child with a set of 11 number cards from − 5 to 5. Children decide who is aiming to collect even differences (2, 4, 6 …), and who is aiming to collect odd differences (1, 3, 5, 7 …). Each child shuffles their set of cards and places them in a pile in front of them, face down. They turn over their top card and together work out the difference. The child who wins that round takes both cards and places them to one side.

The game continues for 11 rounds until each child has used all their cards. Each child then counts how many cards they collected from their winning rounds. The overall winner is the child with more cards.

 ## Counting on and back

Divide the class into two teams. One team is counting on, aiming to finish the game after 10 rounds on a positive number, while the other team is counting back, aiming to finish the game on a negative number.

Using a coin, tell the class that rolling heads means the positive number team counting on and rolling tails means the negative number team counting back.

Start from 0: you or a helper toss the coin and roll a 0–9 or 1–6 dice. The appropriate team then counts on or back the corresponding number of steps and says the number they reach. For example, if you roll tails and 6, the counting back team counts back 6 from 0 to reach − 6.

Starting from − 6, toss the coin and roll the dice again. Once more the appropriate team then counts on or back from that point the corresponding number of steps and says the number they reach.

Keep going like this for 10 rounds, the appropriate team counting on or back from the last number reached.

My record sheet

Name: _____

Date: _____

	Before the activities			After the activities		
I can find the difference between a positive and a negative number.	☺	☺	☹	☺	☺	☹
I can find the difference between two negative numbers.	☺	☺	☹	☺	☺	☹

⬤ After the activities

All these pairs of positive and negative numbers have a difference of 6.	☐ and ☐ ☐ and ☐	☐ and ☐ ☐ and ☐
All these pairs of negative numbers have a difference of 6.	☐ and ☐ ☐ and ☐	☐ and ☐ ☐ and ☐

Individual activity 1A

Name: _____

Date: _____

Find the difference between a positive and
a negative, or two negative integers, in context
[restricted range of numbers]

These thermometers show the temperatures in six different European cities.

Rome
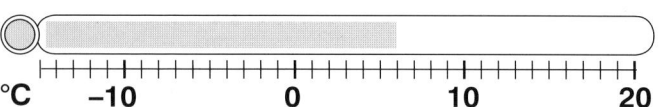

London
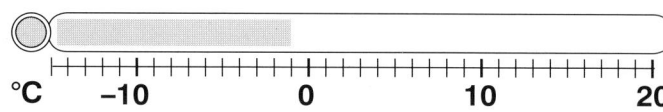

Paris
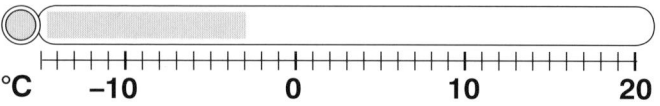

Madrid
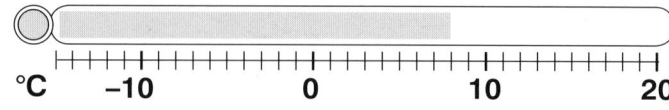

Moscow
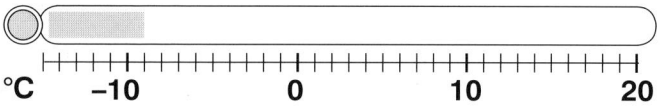

Berlin
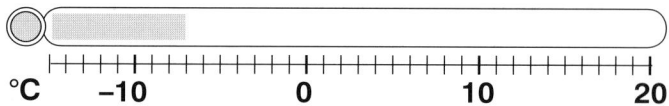

Read the thermometers to work out the differences in temperatures between these cities.

1)   **Rome and London**

2)   **Madrid and Paris**

3)   **Rome and Paris**

4)   **Madrid and London**

5)   **Moscow and London**

6)   **Berlin and Paris**

7)   **Berlin and London**

8)   **Moscow and Berlin**

Now check your answers with your partner.

Using the temperatures on both sheets, take turns to ask each other questions
about the differences in temperatures between different cities.

Answers to 1B

1) 14 °C   2) 11 °C   3) 11 °C   4) 9 °C   5) 1 °C   6) 6 °C   7) 3 °C   8) 4 °C

Individual activity 1B

Name: _____

Date: _____

Find the difference between a positive and a negative, or two negative integers, in context [restricted range of numbers]

These thermometers show the temperatures in six different cities in the USA.

Washington

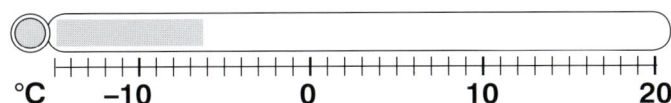

°C   −10   0   10   20

New York

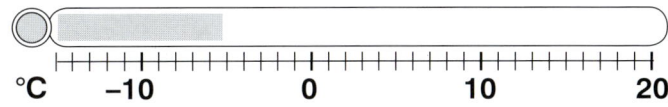

°C   −10   0   10   20

Miami

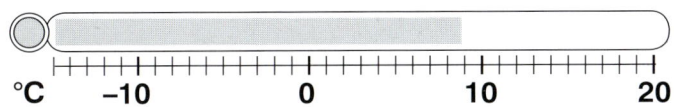

°C   −10   0   10   20

Chicago

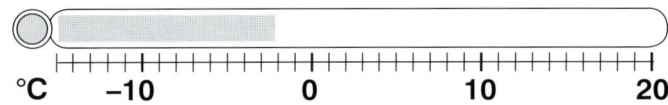

°C   −10   0   10   20

Boston

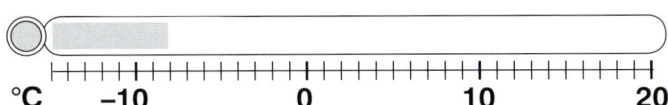

°C   −10   0   10   20

Seattle

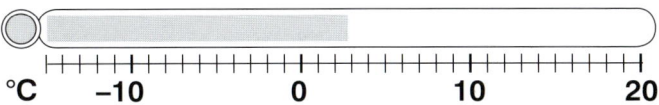

°C   −10   0   10   20

Read the thermometers to work out the differences in temperatures between these cities.

1) Miami and New York

5) Washington and New York

2) Seattle and Boston

6) Chicago and Boston

3) Miami and Chicago

7) New York and Boston

4) Seattle and Washington

8) Chicago and Washington

Now check your answers with your partner.

Using the temperatures on both sheets, take turns to ask each other questions about the differences in temperatures between different cities.

Answers to 1A

1) 7 °C   2) 11 °C   3) 9 °C   4) 9 °C   5) 8 °C   6) 4 °C   7) 6 °C   8) 2 °C

Paired activity 1A

Name: _____

My partner's name: _____

Date: _____

Find the difference between a positive and a negative, or two negative integers, in context [restricted range of numbers]

These thermometers show the temperatures in six different cities.

Your partner has six thermometers showing the temperatures in six other cities.

Work together to find out the differences in temperatures between these cities.

Stockholm

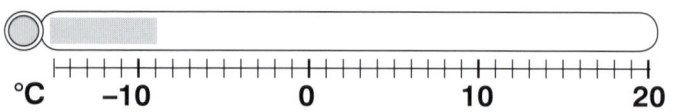

Montreal

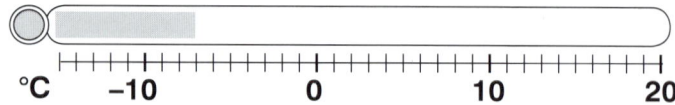

Sydney

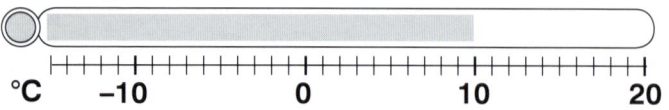

Copenhagen

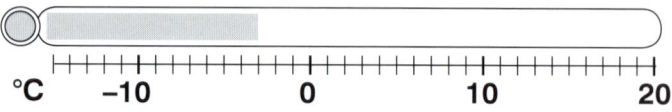

Seoul

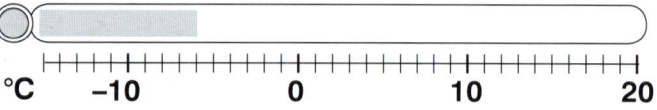

Glasgow

1) Sydney and Tokyo

2) Glasgow and Beijing

3) Seoul and London

4) Stockholm and Cardiff

5) Stockholm and Wellington

6) Copenhagen and Oslo

7) Montreal and Oslo

8) Seoul and Tokyo

Answers to 1B

1) 11 °C 2) 9 °C 3) 12 °C 4) 15 °C 5) 2 °C 6) 2 °C 7) 1 °C 8) 1 °C

Paired activity 1B

Find the difference between a positive
and a negative, or two negative integers,
in context [restricted range of numbers]

Name: _____

My partner's name: _____

Date: _____

These thermometers show the temperatures in six different cities.

Your partner has six thermometers showing the temperatures in six other cities.

Work together to find out the differences in temperatures between these cities.

Tokyo

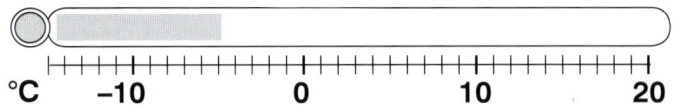

°C   −10   0   10   20

Wellington

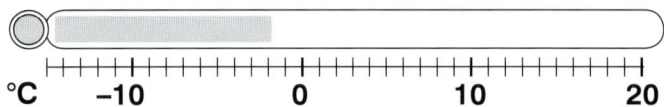

°C   −10   0   10   20

Cardiff

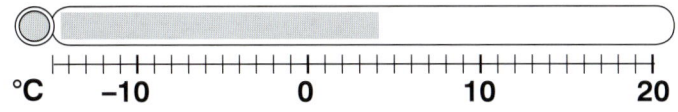

°C   −10   0   10   20

Beijing

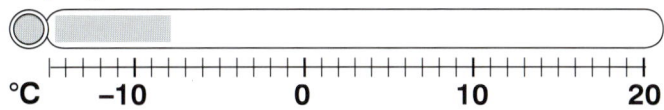

°C   −10   0   10   20

Oslo

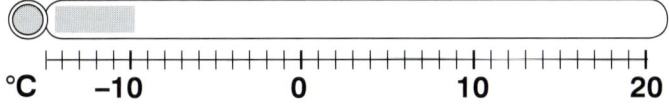

°C   −10   0   10   20

London

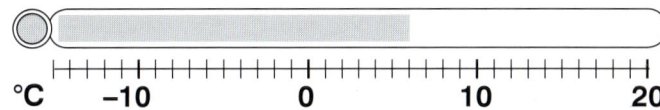

°C   −10   0   10   20

1) **Cardiff and Montreal**

2) **London and
    Copenhagen**

3) **Oslo and Glasgow**

4) **Tokyo and Sydney**

5) **Tokyo and Montreal**

6) **Beijing and Seoul**

7) **Oslo and Stockholm**

8) **Wellington and
    Copenhagen**

Answers to 1A

1) | 15 °C | 2) | 10 °C | 3) | 12 °C | 4) | 13 °C | 5) | 7 °C | 6) | 7 °C | 7) | 3 °C | 8) | 1 °C |

Individual activity 2A

Find the difference between a positive and
a negative, or two negative integers, in context
[wider range of numbers]

These weather charts show the temperatures in six different cities.

Cape Town		24 °C
Helsinki		− 14 °C
Vladivostok		− 33 °C

Vancouver		− 17 °C
Singapore		36 °C
Prague		− 2 °C

Use the weather charts to work out the differences in temperatures between these cities.

1) **Cape Town and Prague**

2) **Singapore and Helsinki**

3) **Cape Town and Vladivostok**

4) **Singapore and Vancouver**

5) **Helsinki and Vancouver**

6) **Prague and Vladivostok**

7) **Vancouver and Prague**

8) **Helsinki and Vladivostok**

Check your answers with your partner.

Using the weather charts on both sheets, take turns to ask each other questions
about the differences in temperatures between different cities.

Answers to 2B

1) 37 °C 2) 68 °C 3) 42 °C 4) 53 °C 5) 5 °C 6) 15 °C 7) 22 °C 8) 12 °C

Individual activity 2B

Name:

Date:

Find the difference between a positive and a negative, or two negative integers, in context [wider range of numbers]

These weather charts show the temperatures in six different cities.

Los Angeles		28 °C
Auckland		− 9 °C
Gorky		− 36 °C

Pyongyang		− 21 °C
Gothenburg		− 14 °C
Nice		32 °C

Use the weather charts to work out the differences in temperatures between these cities.

1) Los Angeles and Auckland

2) Nice and Gorky

3) Los Angeles and Gothenburg

4) Nice and Pyongyang

5) Auckland and Gothenburg

6) Pyongyang and Gorky

7) Gorky and Gothenburg

8) Auckland and Pyongyang

Check your answers with your partner.

Using the weather charts on both sheets, take turns to ask each other questions about the differences in temperatures between different cities.

Answers to 2A

1) 26 °C 2) 50 °C 3) 57 °C 4) 53 °C 5) 3 °C 6) 31 °C 7) 15 °C 8) 19 °C

Paired activity 2A

Name:

My partner's name:

Date:

Find the difference between a positive and a negative, or two negative integers, in context [wider range of numbers]

The chart below shows the summer and winter temperatures for five different European capital cities.

Your partner has a chart that shows the summer and winter temperatures for five other European capital cities.

Capital city	Country	Summer temperature	Winter temperature
Warsaw	Poland	24 °C	− 15 °C
Budapest	Hungary	22 °C	− 11 °C
Dublin	Ireland	27 °C	− 2 °C
Prague	Czech Republic	25 °C	− 8 °C
Brussels	Belgium	28 °C	− 5 °C

Work together to find out the differences in these temperatures.

1) The summer temperature in Warsaw and the winter temperature in Amsterdam

2) The summer temperature in Prague and the winter temperature in Vienna

3) The summer temperature in Amsterdam and the winter temperature in Budapest

4) The summer temperature in Berne and the winter temperature in Prague

5) The winter temperatures in Budapest and Berlin

6) The winter temperatures in Prague and Amsterdam

7) The winter temperatures in Kiev and Dublin

8) The winter temperatures in Vienna and Warsaw

Paired activity 2B

Name:

My partner's name:

Date:

Find the difference between a positive and a negative, or two negative integers, in context [wider range of numbers]

The chart below shows the summer and winter temperatures for five different European capital cities.

Your partner has a chart that shows the summer and winter temperatures for five other European capital cities.

Capital city	Country	Summer temperature	Winter temperature
Kiev	Ukraine	21 °C	− 17 °C
Amsterdam	Netherlands	26 °C	− 2 °C
Vienna	Austria	23 °C	− 8 °C
Berlin	Germany	29 °C	− 3 °C
Berne	Switzerland	19 °C	− 6 °C

Work together to find out the differences in these temperatures.

1) The summer temperature in Kiev and the winter temperature in Brussels

2) The summer temperature in Berlin and the winter temperature in Dublin

3) The summer temperature in Budapest and the winter temperature in Kiev

4) The summer temperature in Dublin and the winter temperature in Berlin

5) The winter temperatures in Berne and Brussels

6) The winter temperatures in Vienna and Budapest

7) The winter temperatures in Prague and Berlin

8) The winter temperatures in Warsaw and Amsterdam

Section 5

Deriving and recalling multiplication and division facts, and using that knowledge

Level 1
- Recall squares of numbers to 12×12

Level 2
- Derive squares of multiples of 10 to 120×120

Strategic approaches to develop fluency in recalling squares of numbers to 12 × 12 and deriving squares of multiples of 10 to 120 × 120

 Representing square numbers as square arrays

Children need to realise that some numbers such as 1, 4, 9, 16, 25, and so on, which can be represented by square arrays are called 'square numbers'.

Arrays can be represented as dots or small

squares. If the children see the arrays as squares, called 'square units', rather than just as dots, then the number of squares in the array also corresponds to the total area: for example, the area of the 5 by 5 square grid is 25 square units.

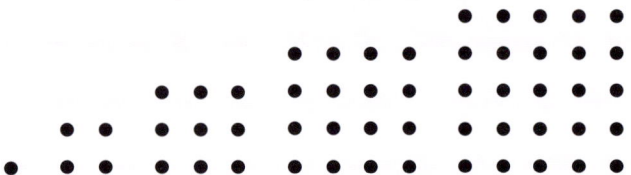

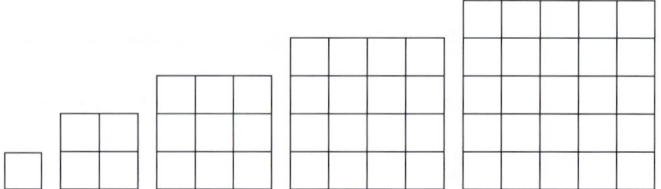

 Calculating square numbers

The notion of arrays explains the geometric idea of square numbers. The arithmetic idea of square numbers is that a square number is the product of multiplying a whole number by itself. In the

above illustration, the number 9 is represented by 3 rows of 3 dots, or 3 rows of 3 squares, which corresponds to the multiplication fact 3 × 3.

 Mathematical notation and language

Children comprehend quite quickly that there is a special mathematical symbol that they can use as shorthand for writing 3 × 3. This is: 3^2, which

means that there are two threes to be multiplied together. You read this as 'three squared' or 'three to the power of two'.

 Using known facts

Squares of multiples of 10 to 120 × 120, such as 30 × 30, are made easier when children use what they know about the squares of numbers to 12 × 12.

It is important that children know their multiplication and division facts to 10 × 10 thoroughly so that they can use these facts to apply to answering related facts.

 Link squares of multiples of 10 to array representations

Open arrays can further help children appreciate the effect of multiplying a multiple of 10 by itself.

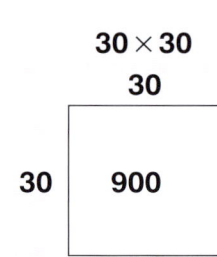

3 × 3

3

3 | 9

30 × 30

30

30 | 900

Individual and paired activities

Level 1 Recall squares of numbers to 12 × 12

The individual activity provides practice in recalling squares of numbers to 12 × 12. Children multiply each number by itself and write the answer in the grey box underneath. They then find the letter that matches each answer to reveal a message.

1A Message

121	144	64	25	121	36	81	49	121	100	144	81	16	
I	N	S	P	I	R	A	T	I	O	N	A	L	!

1B Message

81	49	25	81	36	121	9	64	16	100	144	
E	X	C	E	P	T	I	O	N	A	L	!

What is four squared?

What number multiplied by itself gives 36?

The paired activity provides further practice in recalling squares of numbers to 12 × 12 and also an opportunity for children to give each others instructions and to listen carefully. For the second part of the activity, encourage the children to say each of their numbers to their partner in quick succession. The children will have already found the squares of some of the numbers in the first part of the activity, so should find this relatively easy.

What number did you write in your first box?

What is this number multiplied by itself?

Level 2 Derive squares of multiples of 10 to 120 × 120

In the individual activity, children derive the squares of multiples of 10 to 120 × 120 and colour the answers on a grid. If both children are correct, they should have coloured the same pattern.

What is 30 multiplied by 30? How did you work it out? Did you use another fact to help you? What fact was this?

Explain to me how you work out the squares of multiples of 10 such as 70 times 70.

In the paired activity, children take turns to say numbers from 1 to 12 and multiples of 10 from 10 to 120 to each other. Each child then squares each of the numbers their partner tells them and writes the answer in a box. Pairs compare sheets when they have finished. They should notice that one child's answer is the square of a number from 1 to 12, while the other child's answer is the related squared multiple of 10.

What is 6 times 6? If 6 times 6 is 36, then what is 60 times 60?

What do you notice about each other's answers? Why is this so?

Further activities to develop fluency

 Squared dice

Provide each pair or group with a 0–9 or 1–12 dice and 20 counters. Children take turns to roll the dice. They multiply the number rolled by itself. The first child to call out the correct answer takes a counter. The winner is the player with more counters after 10 rounds.

 Three squared investigations

1. Squared number pattern

What do you notice about this number pattern?

What is the rule? (Answer: square the first digit and add a second digit.)

Investigate what other number patterns you can make using this rule.

$72 \longrightarrow 49 + 2 = 51$

$51 \longrightarrow 25 + 1 = 26$

$26 \longrightarrow 4 + 6 = 10$

$10 \longrightarrow 1 + 0 = 1$

72, 51, 26, 10, 1

2. Another squared number pattern

Children choose a number from 2 to 10 and multiply the number by itself.

They then multiply together the two numbers that are one more and one less than their chosen number.

They repeat for other numbers.

What do you notice?

(Answer: When a number is squared, e.g. $6^2 = 36$, the answer is always one more than the product of the two numbers that are one more and one less than the number, i.e. $5 \times 7 = 35$.)

Does this work for numbers greater than 10?

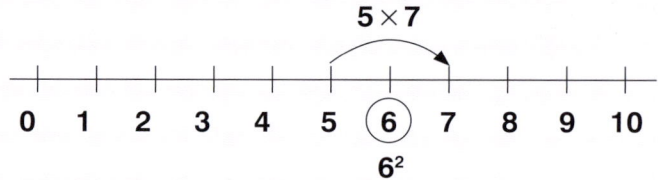

3. Square differences

Investigate which numbers from 1 to 50 you can make by calculating the difference between two square numbers.

What about finding the sum of two (or more) square numbers?

$12 = 4^2 - 2^2$

 Using a calculator

Children investigate a shortcut for obtaining square numbers on a calculator.
(Answer: [3] [×] [=])

My record sheet

Name: _____

Date: _____

	Before the activities	After the activities
I can say the squares of numbers to 12 × 12.	☺ 😐 ☹	☺ 😐 ☹
I can work out the squares of multiples of 10 to 120 × 120.	☺ 😐 ☹	☺ 😐 ☹

 ## After the activities

I have multiplied each of the numbers I have written in the squares by itself and written the answer in the circle.	Write a different number from 2 to 12 in each of the squares. □ ○ □ ○ □ ○
I have multiplied each of the numbers I have written in the squares by itself and written the answer in the circle.	Write a different multiple of 10 from 10 to 120 in each of the squares. □ ○ □ ○ □ ○

Individual activity 1A

Name:

Date:

Recall squares of numbers to 12 × 12

Work out the square of each number and
write the answer in the grey box underneath.

5	7	11	9	12	6	4	8	10
P	T	I	A	N	R	L	S	O

Look at the grid below.
Find the letter that matches the answer in the grey box above.
Write the letter below the number.
What does it read?

121	144	64	25	121	36	81	49	121	100	144	81	16

!

If it doesn't make sense, go back and check your work.

Check with your partner to see if you both have
the same answers for all the squares to 12 × 12.

Individual activity 1B

Recall squares of numbers to 12 × 12

Work out the square of each number and
write the answer in the grey box underneath.

5	11	3	9	12	6	4	7	10	8
C	T	I	E	L	P	N	X	A	O

Look at the grid below.
Find the letter that matches the answer in the grey box above.
Write the letter below the number.
What does it read?

81	49	25	81	36	121	9	64	16	100	144

!

If it doesn't make sense, go back and check your work.

Check with your partner to see if you both have
the same answers for all the squares to 12 × 12.

Paired activity 1A

Recall squares of numbers to 12 × 12

Name:

My partner's name:

Date:

Write any eight numbers from 2 to 12 in any order in the boxes below.
Now multiply each number by itself and write the answer in the circle.

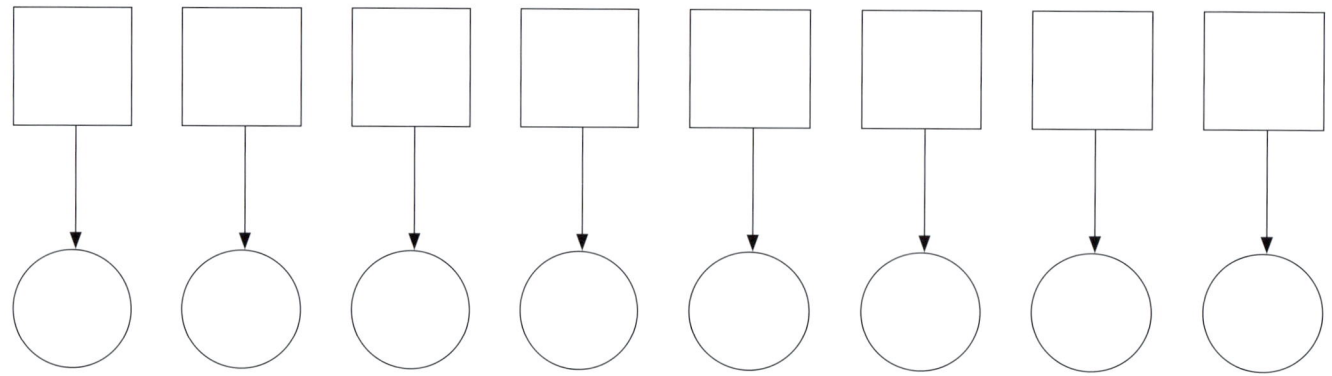

Speak Take turns to say each of the numbers in the boxes above to your partner.

Listen Listen to your partner.
When they say a number, write this in one of the grey boxes below.

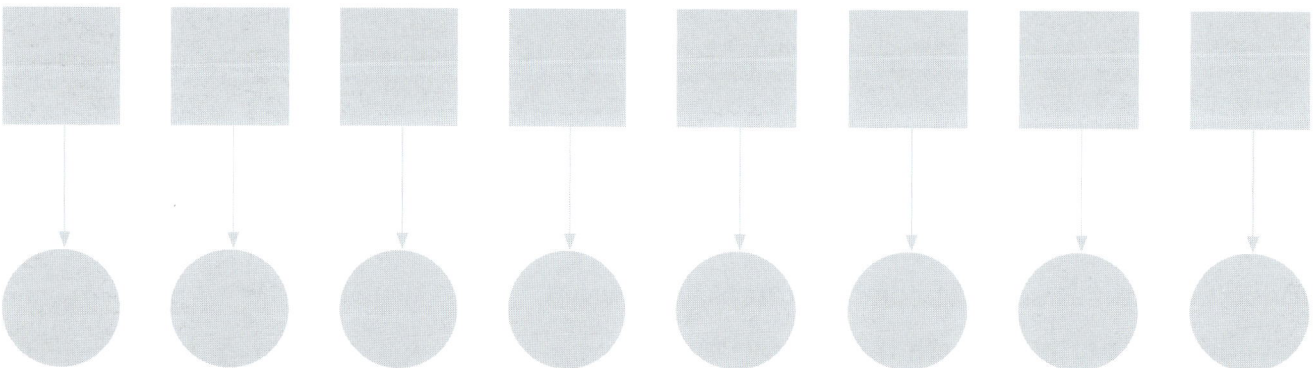

Now multiply each number in the grey boxes above by itself
and write the answer in the grey circle.

Compare your answers with your partner's.

Talk with your partner about any differences in your answers.

Paired activity 1B

Recall squares of numbers to 12 × 12

Name:

My partner's name:

Date:

Write any eight numbers from 2 to 12 in any order in the boxes below.
Now multiply each number by itself and write the answer in the circle.

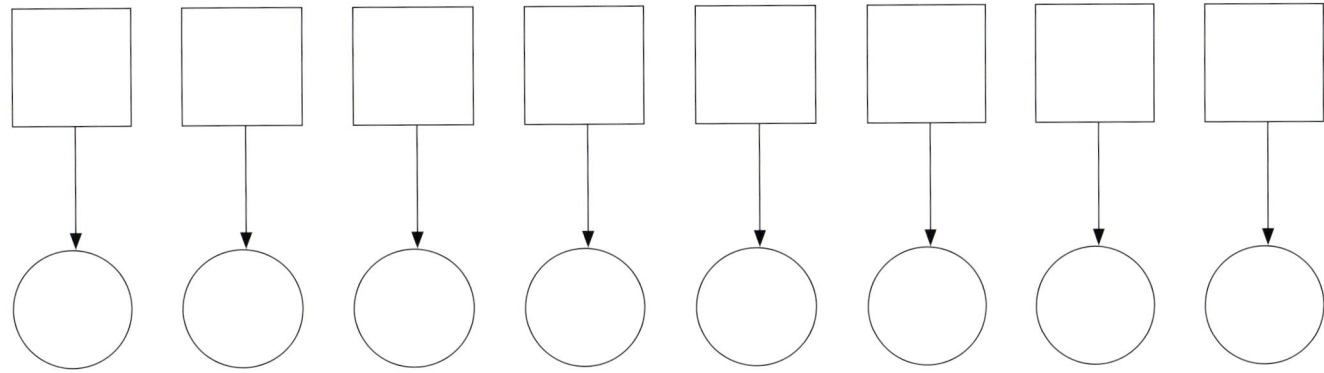

Speak Take turns to say each of the numbers in the boxes above to your partner.

Listen Listen to your partner.
When they say a number, write this in one of the grey boxes below.

Now multiply each number in the grey boxes above by itself
and write the answer in the grey circle.

Compare your answers with your partner's.

Talk with your partner about any differences in your answers.

Individual activity 2A

Derive squares of multiples of 10 to 120 × 120

Work out the answer to each of these multiplication facts.

You need:
• coloured pencil

40 × 40 =

60 × 60 =

10 × 10 =

120 × 120 =

50 × 50 =

70 × 70 =

20 × 20 =

100 × 100 =

90 × 90 =

30 × 30 =

110 × 110 =

80 × 80 =

Look at your answers in the triangles above.

Find each of these numbers on the grid and colour them.

When you have finished, compare your pattern with your partner's pattern.

Have you both drawn the same pattern?

If not find out why?

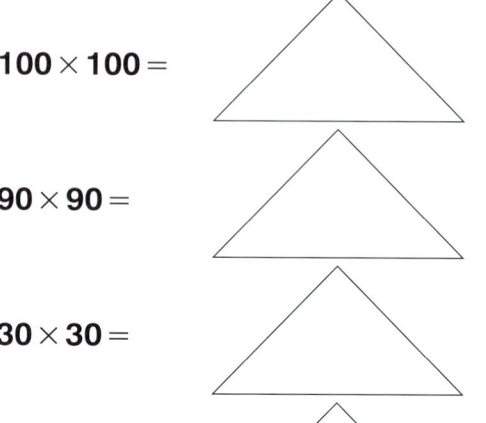

3600		4200	1600		4500
	3200	10 000		2000	100
	8100	5200		8200	6600
600		1200	11 000		4900
900		7200	1400		2200
	1000	2400		3800	14 400
	2500	200		12 100	5400
800		400	4800		6400

Individual activity 2B

Derive squares of multiples of 10 to 120 × 120

You need:
- coloured pencil

Work out the answer to each of these multiplication facts.

50 × 50 =

80 × 80 =

10 × 10 =

120 × 120 =

30 × 30 =

20 × 20 =

100 × 100 =

90 × 90 =

40 × 40 =

60 × 60 =

110 × 110 =

70 × 70 =

Look at your answers in the triangles above.

Find each of these numbers on the grid and colour them.

When you have finished, compare your pattern with your partner's pattern.

Have you both drawn the same pattern?

If not find out why?

4900		2400	8100		3800
	200	1600		5200	900
	12 100	1200		2000	800
1400			7200	4200		2500
10 000			600	11 000		1000
	3200	5400		4800	3600
	100	4500		14 400	6600
2200			6400	8200		400

Paired activity 2A

Derive squares of multiples of 10 to 120 × 120

Name: _____

My partner's name: _____

Date: _____

Speak Take turns to say each of these numbers to your partner.

1) 6   4) 80   7) 120   10) 7

2) 3   5) 2   8) 10   11) 11

3) 100   6) 9   9) 40   12) 50

Listen Listen to the numbers your partner says. Multiply each number by itself and write the answers in the boxes below.

1) [ ]   4) [ ]   7) [ ]   10) [ ]

2) [ ]   5) [ ]   8) [ ]   11) [ ]

3) [ ]   6) [ ]   9) [ ]   12) [ ]

Compare your answers with your partner's.

What do you notice?

Paired activity 2B

Name:

My partner's name:

Date:

Derive squares of multiples of 10 to 120 × 120

Speak Take turns to say each of these numbers to your partner.

1) 60   4) 8   7) 12   10) 70

2) 30   5) 20   8) 1   11) 110

3) 10   6) 90   9) 4   12) 5

Listen Listen to the numbers your partner says. Multiply each number by itself and write the answers in the boxes below.

1) ☐   4) ☐   7) ☐   10) ☐

2) ☐   5) ☐   8) ☐   11) ☐

3) ☐   6) ☐   9) ☐   12) ☐

Compare your answers with your partner's.

What do you notice?

Mental calculation methods

Level 1
- Calculate mentally with integers: specifically,
  TU ± TU, TU × U, TU ÷ U

Level 2
- Calculate mentally with decimals: specifically,
  U.t ± U.t, U.t × U, U.t ÷ U

Strategic approaches to develop fluency in calculating with whole numbers: TU ± TU, TU × U, TU ÷ U; and decimals: U.t ± U.t, U.t × U, U.t ÷ U

 ## Focus on the commutative law

Encourage children not simply to carry out calculations in the order in which they are presented on the page: for example, given $28 + 76$, it is much easier to calculate $76 + 28$.

 ## Use inverse relationships and known facts

Encourage children to think about different ways of answering a calculation before choosing one. For example, you can do a calculation like $93 - 57$ by taking away 57 from 93, but you can also figure it out by asking yourself what you need to add to 57 to make 93 (finding the difference by counting up). You might still decide that 'taking away' 57 is the most sensible approach, but it is a good 'mathematical habit of mind' to explore different approaches rather than going for the first one that comes to mind.

When asked, for example, to calculate $48 \div 6$, fluent calculators know that $8 \times 6 = 48$ and use their knowledge of that multiplication fact to know quickly that $48 \div 6 = 8$.

Encourage children to talk about related multiplication and division facts. For example, there are four related facts using the trio 6, 8, 48; two multiplication and two division:

$$6 \times 8 = 48$$
$$8 \times 6 = 48$$
$$48 \div 6 = 8$$
$$48 \div 8 = 6$$

By firmly embedding the trio 6, 8, 48, all the related multiplication and division facts become fluent.

In the same way, reinforce the division facts corresponding to multiplication facts involving decimals:

$$0.6 \times 8 = 4.8$$
$$0.8 \times 6 = 4.8$$
$$4.8 \div 6 = 0.8$$
$$4.8 \div 8 = 0.6$$

 ## Use knowledge of place value and partitioning

Make sure that children have a firm understanding of place value and can partition numbers with confidence.

Solving multiplication calculations such as 57×4 or 8.3×6 involves partitioning one of the numbers and then multiplying the parts by the other number. This allows you to multiply the tens and units, or units and tenths, separately to form partial products. You then add these together to find the total product.

$$57 \times 4 = (50 \times 4) + (7 \times 4)$$
$$= 200 + 28$$
$$= 228$$

$$8.3 \times 6 = (8 \times 6) + (0.3 \times 6)$$
$$= 48 + 1.8$$
$$= 49.8$$

The use of place value and partitioning also applies to addition and subtraction calculations:

$$63 + 78 = 60 + 3 + 70 + 8$$
$$= 130 + 11$$
$$= 141$$

or

$$63 + 78 = 78 + 60 + 3$$
$$= 138 + 3$$
$$= 141$$

$$4.7 + 9.5 = 4 + 0.7 + 9 + 0.5$$
$$= 13 + 1.2$$
$$= 14.2$$

or

$$4.7 + 9.5 = 9.5 + 4 + 0.7$$
$$= 13.5 + 0.7$$
$$= 14.2$$

$$93 - 57 = 93 - 50 - 7$$
$$= 43 - 7$$
$$= 36$$

$$7.4 - 3.5 = 7.4 - 3 - 0.5$$
$$= 4.4 - 0.5$$
$$= 3.9$$

Individual and paired activities

Level 1 Calculate mentally with integers: specifically, TU ± TU, TU × U, TU ÷ U

In the individual activity, children complete a cross-number puzzle involving the following types of whole number calculations: TU ± TU, TU × U, TU ÷ U

What is 72 divided by 4? How did you work it out?

If you know that there are 8 units in the answer, can this answer be correct? How could you check?

The paired activity is a version of the 'Loop' card game. Children take turns to work out the answer to a whole number calculations (TU ± TU, TU × U, TU ÷ U), tell their partner the answer and then receive a letter from their partner which they use to create a message.

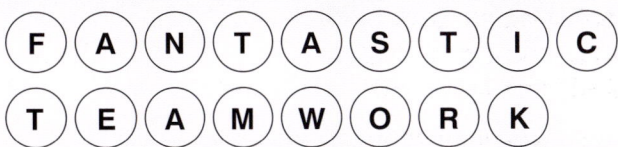

What is the answer to 28 add 34? (62)
What letter goes with the answer? (F)
What is the next calculation? (23 × 4)

Level 2 Calculate mentally with decimals: specifically, U.t ± U.t, U.t × U, U.t ÷ U

In the individual activity, children answer the following types of decimal calculations: U.t ± U.t, U.t × U, U.t ÷ U. They write their answers in boxes and check their answers with their partner.

What is 3.2 divided by 4? How did you work it out? Did you use a different fact to help you work it out? What fact was that?

How did you work out the answer to that calculation?

The paired activity is another version of the 'Loop' card game. This time, children work out the answers to decimal calculations: U.t ± U.t, U.t × U, U.t ÷ U.

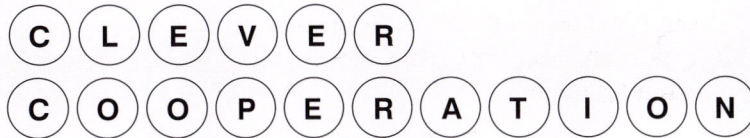

What is the answer to 3.6 plus 5.8? (9.4)
What letter goes with the answer? (C)
What is the next calculation? (7.8 × 3)

Further activities to develop fluency

 ## Addition squares

Display the addition square for numbers 0 to 10 and discuss with the children its features and how to use it, including how to find the related subtraction facts.

Make sure that the children are familiar with this before displaying an addition square for the related facts involving decimals.

+	0	1	2	3	4	5	6	7	8	9	10
0	0	1	2	3	4	5	6	7	8	9	10
1	1	2	3	4	5	6	7	8	9	10	11
2	2	3	4	5	6	7	8	9	10	11	12
3	3	4	5	6	7	8	9	10	11	12	13
4	4	5	6	7	8	9	10	11	12	13	14
5	5	6	7	8	9	10	11	12	13	14	15
6	6	7	8	9	10	11	12	13	14	15	16
7	7	8	9	10	11	12	13	14	15	16	17
8	8	9	10	11	12	13	14	15	16	17	18
9	9	10	11	12	13	14	15	16	17	18	19
10	10	11	12	13	14	15	16	17	18	19	20

+	0.0	0.1	0.2	0.3	0.4	0.5	0.6	0.7	0.8	0.9	1
0.0	0.0	0.1	0.2	0.3	0.4	0.5	0.6	0.7	0.8	0.9	1
0.1	0.1	0.2	0.3	0.4	0.5	0.6	0.7	0.8	0.9	1	1.1
0.2	0.2	0.3	0.4	0.5	0.6	0.7	0.8	0.9	1	1.1	1.2
0.3	0.3	0.4	0.5	0.6	0.7	0.8	0.9	1	1.1	1.2	1.3
0.4	0.4	0.5	0.6	0.7	0.8	0.9	1	1.1	1.2	1.3	1.4
0.5	0.5	0.6	0.7	0.8	0.9	1	1.1	1.2	1.3	1.4	1.5
0.6	0.6	0.7	0.8	0.9	1	1.1	1.2	1.3	1.4	1.5	1.6
0.7	0.7	0.8	0.9	1	1.1	1.2	1.3	1.4	1.5	1.6	1.7
0.8	0.8	0.9	1	1.1	1.2	1.3	1.4	1.5	1.6	1.7	1.8
0.9	0.9	1	1.1	1.2	1.3	1.4	1.5	1.6	1.7	1.8	1.9
1	1	1.1	1.2	1.3	1.4	1.5	1.6	1.7	1.8	1.9	2

Discuss the following with the children:

- If you know that $6 + 8 = 14$, then you can also work out facts such as: $60 + 80 = 140$, $600 + 800 = 1400$, $6000 + 8000 = 14\,000$
- Just as $6 + 8 = 14$, so too does $0.6 + 0.8 = 1.4$
- You can do addition in any order: $0.6 + 0.8 = 1.4$ and $0.8 + 0.6 = 1.4$

- Addition is the inverse of subtraction. If you know that $0.6 + 0.8 = 1.4$, you also know that $1.4 - 0.6 = 0.8$ and $1.4 - 0.8 = 0.6$

See Section 3 for a similar idea related to multiplication and division of whole numbers (and decimals – multiplication squares)

 ## Make a sum

Provide each pair with two 0–9 dice and paper and pencil each. Children take turns to roll the dice: for example, 6 and 3. They use the numbers rolled to make a two-digit number: for example, 63 or 36.

Children roll both dice again, make another two-digit number and add their numbers together. The winner of the round is the player with the larger total. Children play 5 rounds.

Variations:

- Play 'Make a difference'. The winner of the round is the player with the greater difference.
- Play 'Make a decimal sum'. Children use the numbers rolled to make a two-digit decimal: for example, 6.3 or 3.6.
- Play 'Make a decimal difference'.

My record sheet

Name:

Date:

	Before the activities			After the activities		
I can add and subtract numbers such as $54 + 78$ and $84 - 36$ in my head.	☺	😐	☹	☺	😐	☹
I can multiply numbers such as 64×5 in my head.	☺	😐	☹	☺	😐	☹
I can divide numbers such as $91 \div 7$ in my head.	☺	😐	☹	☺	😐	☹
I can add and subtract decimals such as $6.2 + 8.7$ and $9.2 - 5.4$ in my head.	☺	😐	☹	☺	😐	☹
I can multiply decimals such as 3.8×4 in my head.	☺	😐	☹	☺	😐	☹
I can divide numbers such as $4.8 \div 6$ in my head.	☺	😐	☹	☺	😐	☹

 After the activities

This addition and subtraction both have an answer of 62.	☐ + ☐ = **62**	☐ – ☐ = **62**
Here are two more multiplications like 64×5 that I can do in my head.	☐ × ☐ = ◯	☐ × ☐ = ◯
Here are two more divisions like $91 \div 7$ that I can do in my head	☐ ÷ ☐ = ◯	☐ ÷ ☐ = ◯
This addition and subtraction both have an answer of 5.4.	☐ + ☐ = **5.4**	☐ – ☐ = **5.4**
Here are two more multiplications like 3.8×4 that I can do in my head.	☐ × ☐ = ◯	☐ × ☐ = ◯
Here are two more divisions like $4.8 \div 6$ that I can do in my head.	☐ ÷ ☐ = ◯	☐ ÷ ☐ = ◯

Individual activity 1A

Name:

Date:

Calculate mentally with integers: specifically, TU ± TU, TU × U, TU ÷ U

Work out the answer to each of these calculations and write the number in the correct place on the cross-number puzzle.

Across

1) $72 \div 4$
2) 87×5
5) $84 + 28$
8) $91 - 53$
11) $82 - 28$
12) 23×4

Down

1) $64 + 67$
3) $94 - 38$
4) $28 + 34$
6) 24×7
7) $80 \div 5$
9) $72 \div 3$
10) $95 + 95$

Check your cross-number puzzle with your partner.

Answers to 1B

1) 1	6		2) 5	9	3) 2
4		4) 3			7
5) 2	6) 2	8		7) 2	
	6			6	
8) 8	8		9) 2		10) 1
		11) 1	8		5
12) 1	9				1

Individual activity 1B

Name:

Date:

Calculate mentally with integers:
specifically, TU ± TU, TU × U, TU ÷ U

Work out the answer to each of these calculations and write the number
in the correct place on the cross-number puzzle.

Across

1) $96 \div 6$
2) 74×8
5) 38×6
8) $59 + 29$
11) $90 \div 5$
12) $52 - 33$

Down

1) $79 + 63$
3) $81 \div 3$
4) $74 - 36$
6) 67×4
7) $81 - 55$
9) $66 - 38$
10) $84 + 67$

Check your cross-number puzzle with your partner.

Answers to 1A

1) 1	8		2) 4	3	3) 5
3		4) 6			6
5) 1	6) 1	2		7) 1	
	6			6	
8) 3	8		9) 2		10) 1
		11) 5	4		9
12) 9	2				0

Paired activity 1A

Name: _____

My partner's name: _____

Date: _____

Calculate mentally with integers:
specifically, TU ± TU, TU × U, TU ÷ U

Start at the grey box and work out the answer to the calculation.

Speak Say the answer to your partner.

Listen Your partner will tell you a letter.
Write this letter in the first circle below.

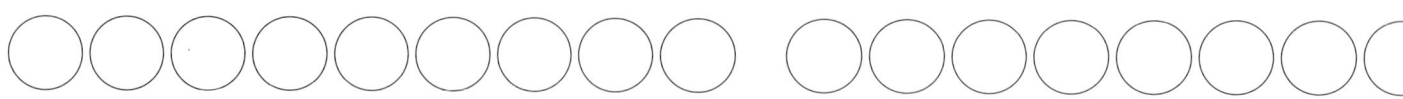

Listen Your partner will now tell you a number.

Find the box that has this number at the top and say the letter
that is in the box to your partner.
Write this letter in the next circle.

Speak Look at the calculation that is at the bottom of the box.
Work out the answer and say the answer to your partner.

Take turns to work out the answer to a calculation, say the answer
to your partner and write the letter your partner tells you in the next circle.

Can you read the message?

28 + 34	14 (W) 42 × 7	99 (T) 34 × 6
27 (I) 51 − 23	134 (A) 81 − 46	38 (R) 77 + 56
47 (S) 27 + 64	180 (T) 72 ÷ 4	92 (A) 96 ÷ 8

Paired activity 1B

Name: _____

My partner's name: _____

Date: _____

Calculate mentally with integers:
specifically, TU ± TU, TU × U, TU ÷ U

Listen Your partner will tell you a number.

Find the box that has this number at the top and say the letter
that is in the box to your partner.

Write this letter in the first circle below.

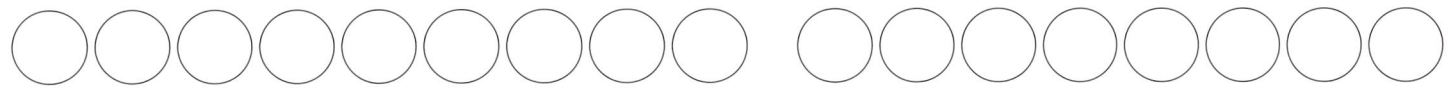

Speak Look at the calculation that is at the bottom of the box.
Work out the answer and say the answer to your partner.

Listen Your partner will tell you a letter.
Write this letter in the next circle.

Take turns to work out the answer to a calculation, say the answer
to your partner and write the letter your partner tells you in the next circle.

Can you read the message?

91 (T) 81 ÷ 3	18 (E) 76 + 58	28 (C) 36 × 5
133 (K)	62 (F) 23 × 4	294 (O) 62 − 24
12 (N) 53 + 46	35 (M) 84 ÷ 6	204 (A) 83 − 36

Individual activity 2A

Name:

Date:

Calculate mentally with decimals:
specifically, **U.t ± U.t, U.t × U, U.t ÷ U**

Complete these calculations.

1) $5.6 + 7.5 =$ ☐

2) $5.8 \times 6 =$ ☐

3) $5.4 \div 6 =$ ☐

4) $6.8 + 9.4 =$ ☐

5) $8.3 - 2.7 =$ ☐

6) $3.2 \div 4 =$ ☐

7) $6.2 \times 7 =$ ☐

8) $7.5 - 3.8 =$ ☐

9) $5.6 \div 8 =$ ☐

10) $8.4 \times 9 =$ ☐

11) $9.2 - 4.5 =$ ☐

12) $4.7 + 8.8 =$ ☐

Write your answers in the grid.

1)	2)	3)	4)	5)	6)	7)	8)	9)	10)	11)	12)

Check your grid with your partner's grid at the bottom of their sheet.

Talk with your partner about how you worked out the answer to some of your calculations.

What facts that you knew by heart did you use to help you?

Answers to 2B

1)	2)	3)	4)	5)	6)	7)	8)	9)	10)	11)	12)
15.9	26.8	18.3	0.9	3.8	39.2	38.7	5.5	0.8	9.4	6.8	0.7

Individual activity 2B

Name:

Date:

Calculate mentally with decimals:
specifically, U.t ± U.t, U.t × U, U.t ÷ U

Complete these calculations.

1) $7.5 + 8.4 =$ ☐

2) $6.7 \times 4 =$ ☐

3) $8.7 + 9.6 =$ ☐

4) $2.7 \div 3 =$ ☐

5) $9.1 - 5.3 =$ ☐

6) $5.6 \times 7 =$ ☐

7) $4.3 \times 9 =$ ☐

8) $7.3 - 1.8 =$ ☐

9) $7.2 \div 9 =$ ☐

10) $5.8 + 3.6 =$ ☐

11) $9.2 - 2.4 =$ ☐

12) $4.2 \div 6 =$ ☐

Write your answers in the grid.

1)	2)	3)	4)	5)	6)	7)	8)	9)	10)	11)	12)

Check your grid with your partner's grid at the bottom of their sheet.

Talk with your partner about how you worked out the answer to some of your calculations.

What facts that you knew by heart did you use to help you?

Answers to 2A

1)	2)	3)	4)	5)	6)	7)	8)	9)	10)	11)	12)
13.1	34.8	0.9	16.2	5.6	0.8	43.4	3.7	0.7	75.6	4.7	13.5

Paired activity 2A

Name: _____

My partner's name: _____

Date: _____

Calculate mentally with decimals:
specifically, U.t ± U.t, U.t × U, U.t ÷ U

Start at the grey box and work out the answer to the calculation.

Speak Say the answer to your partner.

Listen Your partner will tell you a letter.
Write this letter in the first circle below.

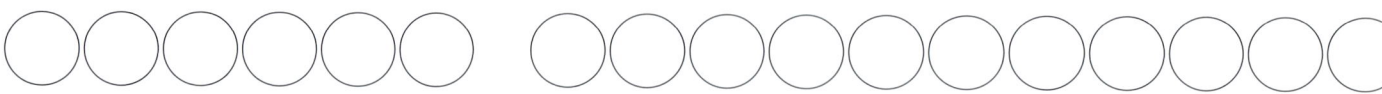

Listen Your partner will now tell you a number.

Find the box that has this number at the top and say the letter
that is in the box to your partner.
Write this letter in the next circle.

Speak Look at the calculation that is at the bottom of the box.
Work out the answer and say the answer to your partner.

Take turns to work out the answer to a calculation, say the answer
to your partner and write the letter your partner tells you in the next circle.

Can you read the message?

	2.4 (V) 9.4 − 3.7	4.2 (R) 7.3 − 5.5
3.6 + 5.8		
11.3 (O) 3.7 × 8	0.9 (T) 4.8 ÷ 8	23.4 (L) 6.3 ÷ 9
0.8 (R) 7.6 × 4	12.4 (O) 5.3 + 7.8	3.6 (P) 8.4 + 5.9

Paired activity 2B

Name: _____

My partner's name: _____

Date: _____

Calculate mentally with decimals:
specifically, U.t ± U.t, U.t × U, U.t ÷ U

Listen Your partner will tell you a number.

Find the box that has this number at the top and say the letter
that is in the box to your partner.

Write this letter in the first circle below.

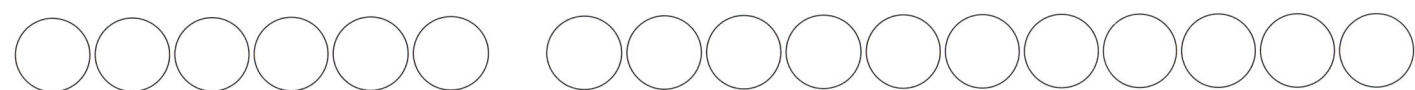

Speak Look at the calculation that is at the bottom of the box.
Work out the answer and say the answer to your partner.

Listen Your partner will now tell you a letter.
Write this letter in the next circle.

Take turns to work out the answer to a calculation, say the answer
to your partner and write the letter your partner tells you in the next circle.

Can you read the message?

0.6 (I) 8.9 + 3.5	29.6 (O) 8.4 − 4.8	30.4 (C) 4.7 + 6.6
5.7 (E) 4.8 ÷ 6	13.1 (N)	14.3 (E) 0.6 × 7
9.4 (C) 7.8 × 3	1.8 (A) 6.3 ÷ 7	0.7 (E) 6.2 − 3.8